Equivalence and Noninferiority Tests for Quality, Manufacturing and Test Engineers

Equivalence and Noninferiority Tests for Quality, Manufacturing and Test Engineers

Scott Pardo

CRC Press
Taylor & Francis Group
Boca Raton London New York

CRC Press is an imprint of the
Taylor & Francis Group, an **informa** business
A CHAPMAN & HALL BOOK

CRC Press
Taylor & Francis Group
6000 Broken Sound Parkway NW, Suite 300
Boca Raton, FL 33487-2742

First issued in paperback 2019

ISBN-13: 978-1-4665-8688-8 (hbk)
ISBN-13: 978-0-367-37945-2 (pbk)

Visit the Taylor & Francis Web site at
http://www.taylorandfrancis.com

and the CRC Press Web site at
http://www.crcpress.com

Contents

Preface

There is much literature on equivalence tests in the clinical bioavailability world. Unfortunately, this paradigm and its associated statistical methods presume that the "limits of equivalence" are actually undesirable values for the variables or parameters under consideration. Thus, for example, if a government regulation states that, on the average, a generic version of a drug must have a time to maximum blood concentration that is within ±20 percent of that for the brand name drug, the regulation actually implies that the difference in those averages must be much less than 20 percent. Thus, statistical equivalence tests for such bioavailability parameters are constructed to FAIL with near certainty (95 percent probability) if the actual difference is 20 percent. An engineer is not likely to specify tolerances or limits on a parameter that are actually unacceptable. The engineer would more likely specify the tolerance around a parameter to be such that fit, form, and function of the product or system or process are not adversely affected as long as the parameter remains within the tolerance limits. So, the methods for "bioequivalence," as they are described by bioavailability writers and researchers, would be useless to the engineer. This work is intended to be a reference handbook for performing analyses related to demonstrating equivalence or noninferiority in the context of engineering and applied science.

The need for equivalence and noninferiority tests in engineering and quality control often arises in the context of process validation and design verification testing. However, there are a wide variety of situations in which equivalence or noninferiority questions may apply. Thus, a variety of tests are presented in this book, with the hope that the practitioner may find a test fitting her or his application, or perhaps gain some insights into an application that may be similar to some of those presented here. The tests described here all stem from actual situations and problems the author has faced in a wide variety of industrial settings over more than 30 years.

In order to get value from this book, the reader should have familiarity with the notions of statistical inference as taught in introductory statistics courses. Test 2.5 and Chapter 7 require exposure to linear regression, including multiple regression. Test 2.6.2 presumes the reader is familiar with two-level factorial experimental designs and analyses. Knowledge of elementary calculus and probability theory would be helpful for understanding the power curve derivations. Knowledge of some matrix algebra is necessary for the multivariate test, Test 6.1. While code and procedures for implementing tests are presented in the "Computational Considerations" sections, this is not a tutorial text on computer programming or software packages. It is presumed that the reader will either be familiar with the software used to implement the tests in the book, or be familiar with another suitable software system.

Finally, this work is heavily invested in the frequentist approach to inference. There are no Bayesian concepts introduced or discussed. This is not because the author is opposed to Bayesian ideas, but because in the context of this work, the frequentist view is more straightforward.

The Author

Scott Pardo has been a professional statistician since 1980. He has worked in a wide variety of industrial contexts, including the U.S. Army Information Systems Command, satellite systems engineering, pharmaceutical development, and medical devices. He is a Six Sigma Master Black Belt, an Accredited Professional Statistician (PStat™), and holds a Ph.D. in Industrial Engineering from the University of Southern California.

Acknowledgments

A number of people have contributed to this work. I would like to thank Mr. Bern Harrison, Dr. John Creaven, and Mr. Mitch Martinez, for their discussions and insights into equivalence testing. Also, thanks are due to Dr. Rezi Zawadzki and Ms. Sarah Kalicin for their encouragement and input. I owe a great debt to Dr. David Grubbs, Ms. Kathryn Everett, and all the staff at Taylor & Francis, for their diligence and editorial guidance. In addition, I would like to thank Dr. Joan L. Parkes for her guidance, leadership, and support. Finally, and most importantly, I owe everything to my wife, Gail, for her encouragement, analytic process, ideas, suggestions, and support, and to my children, Michael, Yehudah, and Jeremy Pardo, for their scientific perspectives and ideas.

Introduction to Hypothesis Testing

Hypothesis testing is the fundamental process of statistical inference. It is in fact the backbone of the scientific method. Unfortunately, it is also the most misunderstood, misapplied, and misinterpreted of all statistical concepts. Aside from the mathematical formalism of hypothesis tests, there are several ideas that make hypothesis testing the central component of empirical investigation.

1. Data are based on a subset (sample) of individuals from a larger, often infinite population.

2. The measurements or quantified observations made on the individuals are associated with a probability distribution, which has some parameters such as the mean, standard deviation, and percentiles.

3. The only information we have (presuming no a priori knowledge of the parameter) about the values of these parameters comes from the sample statistics calculated using the measurements or observations made on individual members of the sample, such as the sample mean, the sample standard deviation, and the sample percentiles.

4. If we obtained two samples from the same population, each sample having the same number of individuals, and computed the sample statistics for some measurement, the two samples would very likely yield different values for the same statistic. Neither sample is likely to yield the actual value of the corresponding population parameter.

5. The number of individuals in the population is too large to feasibly observe every single individual; in other words, we MUST rely on sample information to determine anything about the population parameters.

The most important idea that makes hypothesis testing so indispensable is

6. It is desirable to demonstrate, using empirical observations, that some parameter or set of parameters falls into some particular range of values.

Because we understand from item number 4 that any sample statistic is subject to sampling variation, we must couch our conclusions about population parameters in the language of probability. Furthermore, we must be able to state a priori into what range we are hoping to demonstrate the parameters fall.

Formal Hypothesis Testing

Every hypothesis test begins with a statement of two hypotheses, or asser-
tions, about some parameter or set of parameters that are the logical negations
of each other. The first, by convention, is often called the "null" hypothesis
and is labeled H_0. The alternative, which is the logical negation of the null
hypothesis, is symbolized in this work by H_1. The null hypothesis is generally a statement of something we hope is not true. The idea is that, using
data, we hope to reject the null hypothesis. We reject the null hypothesis if
it is not likely enough to have obtained the observations we made given that
the null hypothesis is true. "Likely enough" is customarily defined as a probability greater than 5 percent. Typically, the null hypothesis is stated in such
a way as to make it possible to assess the limits to a feasible range for sample
statistics given that the null hypothesis is actually true. Note that in elementary statistics texts the null hypothesis is almost always stated as an equality,
either the value of a parameter between two or more groups, or the equality
of a parameter for a single population to a particular value. In the case of
equivalence and noninferiority, the null is usually stated as an inequality,
the nature of which is always adverse to the experimenter's general desire.
    As an example, consider the hypotheses

$$H_0: \mu = 10$$

and its logical negation:

$$H_1: \mu \neq 10$$

Here, the Greek letter μ symbolizes the mean value of some measurement
made on individuals, averaged over all individuals in the population. Note
that it is not possible to calculate the arithmetic average of any measurement for a population with an infinite number of members, and it may not
be economically feasible for a population with a finite number of members.
Without elaborating, let us just assume that some population mean exists,
but its value is unknown. Suppose further that we sample $n = 15$ individuals,
and compute the sample mean and the sample standard deviation:

$$\bar{X} = 9.3$$

$$S = 1.7$$

So, do we believe that the population mean is in fact 10 or not? After all, 9.3
doesn't seem too far off from 10. Perhaps if we sampled again, we might get
a different sample mean that was greater than 10. Perhaps we cannot afford
to sample more than these 15 individuals. Statistical theory tells us how to

quantify the uncertainty in deciding which hypothesis to believe. We can use the fact that if the null hypothesis, H_0, is in fact true, then we would expect only so much deviation between the sample mean and the population mean. Specifically, we would expect the quantity:

$$\frac{\bar{X}-10}{S/\sqrt{n}}$$

to fall between the limits:

$$t_{0.025}(15-1) \approx -2.145$$

and

$$t_{0.975}(15-1) \approx +2.145$$

about 95 percent of the time that we sample $n = 15$ individuals from this population and make the same measurements. The values

$$t_{0.025}(15-1)$$

and

$$t_{0.975}(15-1)$$

are the 2.5th and 97.5th percentiles (respectively) of a student's t distribution with $15 - 1 = 14$ degrees of freedom.

The actual computation is

$$\frac{\bar{X}-10}{S/\sqrt{n}} = \frac{9.3-10}{1.7/\sqrt{15}} \approx -1.595.$$

Since the sample computation (−1.595) lies in between ±2.145, we decide not to reject the idea that the null hypothesis is true. We have set our limit of believability to 95 percent. The sample result falls within the 95 percent believability range.

Suppose instead:

$$\bar{X} = 9.0.$$

Then

$$\frac{\bar{X}-10}{S/\sqrt{n}} = \frac{9.0-10}{1.7/\sqrt{15}} \approx -2.278 < -2.145.$$

Therefore, in this case, we would reject the null hypothesis. Note that rejecting or failing to reject the null hypothesis depends on $\bar{X}$, S, n, and μ. Of course, it also depends on the value of the probability measure for the range of "believability," which will always be set at 95 percent in this work. Furthermore, generally we base the rule for rejecting the null hypothesis on the condition that the null hypothesis is actually true. Thus, we typically choose a rule that would give us a relatively small chance of rejecting the null if in fact it were true. The probability that the sample statistic lies outside the range of believability is called the size of the test, size of the critical region, Type I error risk, or significance level.

Hypotheses for Noninferiority and Equivalence

Equivalence and noninferiority tests are based on the desire to show that something is close enough to ideal to be *acceptable*. Boundary conditions (either lower and upper, in the case of equivalence, or one-sided, in the case of noninferiority) are established for a parameter or set of parameters. The test, based on the sample data, is used to decide whether we should believe that the parameter(s) lies(lie) outside the range of "equivalence" or "noninferiority" (null hypothesis) or whether it is better to believe that the parameter(s) is(are) within the range of equivalence or noninferiority (alternative hypothesis). Unlike the traditional or more common hypothesis test, where the null hypothesis is that the parameter is exactly equal to some specified value (often 0), the equivalence and noninferiority test provides evidence that the parameter is in fact "close enough" to ideal. That is, failure to reject the null hypothesis is not sufficient evidence to truly accept it; all we can say is that we do not have sufficient evidence to reject it. The equivalence test is designed to allow the experimenter to make a much stronger statement about the parameters.

In the more common, traditional hypothesis test, the null hypothesis usually states that at least one parameter of interest equals some specific value. The alternative hypothesis can either be one-sided (parameters are less than or greater than the specified value) or two-sided (the parameters are not equal to the specified value, with no specification about being less than or greater than). In a noninferiority hypothesis, the null and the alternative hypotheses specify whether the parameters are less than or greater than specific values. For example, it may be desirable for the mean response under a new treatment to be no worse than some specified quantity less than the mean response using a currently accepted treatment. As an applied physical science example, suppose a material used to make a plastic part has a high tensile strength. A new material is much less expensive and would be acceptable if it had a tensile strength that was no worse than 10 pounds per square inch (lbs/in²) less than the currently used material. Symbolically, the hypotheses could be expressed as:

$$H_0: \mu_{new} - \mu_{current} < -10.00$$

$$H_1: \mu_{new} - \mu_{current} \geq -10.00$$

One difficulty is that there are now an infinite number of values of μ_{new} $- \mu_{current}$ that would make the null hypothesis true. In the traditional hypothesis test, there is usually exactly one value of the parameter that makes the null hypothesis true. That is, the hypotheses might be

$$H_0: \mu_{new} - \mu_{current} = -10.00$$

$$H_1: \mu_{new} - \mu_{current} > -10.00$$

The only condition that makes the null hypothesis true is $\mu_{new} - \mu_{old} = -10.00$. Thus, in the traditional test, it is relatively straightforward to find a critical region of size α. Recall that α is the probability of rejecting the null hypothesis when it is true. The critical value of this test would be

$$\bar{D}_c = -10.00 + t_{1-\alpha}SE$$

where $t_{1-\alpha}$ is the $100(1-\alpha)$ percentile of a t-distribution with $n_{new} + n_{current} - 2$ degrees of freedom, and

$$SE = \sqrt{\frac{S_{new}^2}{n_{new}} + \frac{S_{current}^2}{n_{current}}}.$$

This critical value represents the *highest believable value* that $\bar{D}$ could have if it were true that the average tensile strength of the new material was in fact only 10 lbs/in² less than the average tensile strength of the current material. So, the null hypothesis would be rejected if the difference in sample means

$$\bar{D} = \bar{X}_{new} - \bar{X}_{current}$$

was greater than the critical value (it is a one-sided test, since the alternative is ">"). Equivalently, the null hypothesis would be rejected if

$$\bar{D} - t_{1-\alpha}SE > -10.00.$$

In this traditional hypothesis test, if $\bar{D}$ were less than the critical value, all we could infer is that we do not have sufficient reason to reject the null hypothesis. So, for example, if $\bar{D} = -7.7$, with $n_{new} = n_{current} = 10$, and $\alpha = 0.05$, $S_{new} = S_{current} = 3.0$,

$$\bar{D} - t_{1-\alpha}SE \approx -7.7 - 1.734\sqrt{\frac{3.0^2}{10} + \frac{3.0^2}{10}} \approx -10.03 < -10.00.$$

Therefore, we would fail to reject the null hypothesis.

In the case of the noninferiority test, the probability of falsely rejecting the null hypothesis depends on how false it actually is. So, if $\mu_{new} - \mu_{old} = -10.10$, you would expect a lower chance of rejecting the null hypothesis than if the truth was that $\mu_{new} - \mu_{old} = -20.00$. Nevertheless, the critical value is still based on the condition that $\mu_{new} - \mu_{old} = -10.00$. However, we now find the sample result that would be the *lowest believable value* for $\bar{D}$ if $\mu_{new} - \mu_{old} = -10.00$. Since we are hoping for a difference greater than -10.00, we would use a critical value that is the lowest we would expect for a difference in sample means if $\mu_{new} - \mu_{old} = -10.00$. Using the notation $t(n_{new} + n_{current} - 2)$ to represent a random variable with a (central) t distribution having $n_{new} + n_{current} - 2$ degrees of freedom, we would want a critical value, $\bar{D}_c$, to be such that

$$\Pr\{\bar{D} \geq \bar{D}_c \mid \mu_{new} - \mu_{cuurent} = -10.00\} = 1 - \beta$$

or equivalently,

$$\Pr\{\bar{D} < \bar{D}_c \mid \mu_{new} - \mu_{cuurent} = -10.00\} = \beta.$$

Thus,

$$\Pr\left\{\frac{\bar{D} - (-10.00)}{SE} < \frac{\bar{D}_c - (-10.00)}{SE} \mid \mu_{new} - \mu_{cuurent} = -10.00\right\} = \beta \Rightarrow$$

$$\Pr\left\{t(n_{new} + n_{current} - 2) < \frac{\bar{D}_c - (-10.00)}{SE} \mid \mu_{new} - \mu_{cuurent} = -10.00\right\} = \beta \Rightarrow$$

$$\frac{\bar{D}_c + 10.00}{SE} = -t_{1-\beta} \Rightarrow$$

$$\bar{D}_c = -10.00 - t_{1-\beta}SE$$

and we would reject the null "inferiority" hypothesis:

$$H_0: \mu_{new} - \mu_{current} < -10.00$$

in favor of "noninferiority" if

$$\bar{D} \geq -10.00 - t_{1-\beta} SE$$

or in other words, if

$$\bar{D} + t_{1-\beta} SE \geq -10.00.$$

Using the same data as the example with the traditional hypothesis, with $\beta = 0.05$,

$$\bar{D} + t_{1-\beta} SE \approx -7.7 + 1.734 \sqrt{\frac{3.0^2}{10} + \frac{3.0^2}{10}} \approx -5.37 > -10.00.$$

So, the null "inferiority" hypothesis would be rejected in favor of noninferiority.

In order to reject the traditional null, assuming the same sample sizes and standard deviations, the difference in sample means would have had to be at least −7.67. Conversely, in order to reject the noninferiority null, the difference in the sample means would have had to be at least −12.32.

The Two Paradigms

There are two paradigms under which equivalence and noninferiority tests are formulated: the bioavailability and quality engineering paradigms. In bioavailability studies, the decision rule is designed so that failure to reject the null hypothesis (nonequivalence or inferiority) when the parameter is exactly equal to the boundary condition is low (Wellek, 2003; Berger and Hsu, 1996; Anderson and Hauck, 1983). This paradigm follows the concept alluded to earlier, in which we choose a rule for rejecting the null based on it being true. Thus, if we decide that the mean concentration of a drug must be between L and U in order for it to be effective, and if the mean concentration is either exactly equal to L or U, the test will only have a small chance, α, of rejecting the null, that the mean concentration is either less than L or greater than U.

The second paradigm is that of quality engineering. In this paradigm, tests to determine whether to accept or reject products or processes, referred to as "acceptance sampling tests," can be formulated as equivalence or noninferi-orty hypothesis tests. As in the case of bioavailability, the null hypothesis is that the parameter is outside of some predefined interval of "equivalence" or "noninferiority," and the alternative is that the parameter is inside the inter-val. However, in an acceptance sampling test the decision rule is designed so that the chance of rejecting the null hypothesis when the parameter is exactly equal to the boundary condition is high (Cowden, 1957; Duncan, 1965; Grant and Leavenworth, 1980). Or alternatively, the boundary conditions are those that are the worst "acceptable" values for the parameter of interest. In this work, we have taken the quality engineering point of view. Thus, we would only want at most a 100β percent chance of failing to reject the null if in fact the parameter exactly equaled the "boundary" value, where β is a relatively small probability. Generally, we will set $\beta = 0.05$. It represents the chance of failing to reject the null hypothesis when it is in fact false, and is referred to as Type II risk. In the quality engineering paradigm, the null is considered false if the parameter actually equals a boundary condition.

Schilling and Neubauer (2009), following Freund (1957), point out that in the acceptance sampling paradigm, there should be a $1 - \alpha$ (α is usually set to a small value, such as 0.05) chance of inferring "acceptability" if the parameter under study is exactly equal to the boundary value for equiva-lence (which they call APL, acceptable parameter level, or alternatively, AQL, acceptable quality level). Duncan (1965, p. 268) describes an example where the critical value of a sample mean is the lowest believable value for the sample mean under the hypothesis that the population mean is equal to the lower boundary for noninferiority. He postulates that a chalk company desires a minimum average density for their chalk of $L = 0.1350$ g/mL. He assumed that σ was known to be 0.006 g/mL. Using a critical region of size 0.05, he found a critical value for the sample mean density with sample size n to be such that:

$$\frac{\bar{X}_c - L}{\sigma/\sqrt{n}} = \frac{\bar{X}_c - 0.1350}{0.006/\sqrt{n}} = z_{0.05} \approx -1.645 \Rightarrow \bar{X}_c = L - z_{0.95}\frac{\sigma}{\sqrt{n}}$$

The hypotheses are

$$H_0\colon \mu_{density} < L$$

and

$$H_1\colon \mu_{density} \geq L.$$

Reject H_0 if

$$\bar{X} \geq \bar{X}_c = L - z_{0.95}\frac{\sigma}{\sqrt{n}}$$

or in other words,

$$\bar{X} + z_{0.95}\frac{\sigma}{\sqrt{n}} \geq L.$$

Thus, under the condition that $\mu_{\text{density}} = L$, there would be a $100(1-\beta)$ percent $=$ 95 percent chance of rejecting the null using Duncan's criterion. So, for example, if $\bar{X} = 0.1349$, and $n = 9$, then

$$\bar{X} + z_{0.95}\frac{\sigma}{\sqrt{n}} \approx 0.1349 + 1.645\frac{0.006}{\sqrt{9}} \approx 0.1382 \geq L = 0.1350.$$

Therefore, we would reject the null hypothesis H_0 in favor of the alternative H_1, even though the sample mean was less than L. Of course, if σ were unknown, replace $z_{0.95}$ with $t_{0.95}$, and σ with s.

In the bioavailability paradigm, the critical value would have been

$$\bar{X}_c = L + z_{0.95}\frac{\sigma}{\sqrt{n}} \approx 0.1350 + 1.645\frac{0.006}{\sqrt{9}} \approx 0.1383.$$

And the null would be rejected only if $\bar{X} > \bar{X}_c$. Since $\bar{X} = 0.1349 \leq 0.1383$, we would have failed to reject H_0.

As another example, consider a binomial proportion, P. The hypotheses are

$$H_0: P < P_0$$

$$H_1: P \geq P_0.$$

In other words, we are hoping that P is greater than or equal to P_0.

In the quality engineering paradigm, we would choose a critical value, P_c, so that if $P = P_0$ exactly, we would expect the sample proportion

$$\hat{P} = \frac{X}{n}$$

to be at least

$$P_c = \frac{X_c}{n}$$

about $1 - \beta \approx 95$ percent of the time, or more exactly, choose X_c over all possible values of X such that:

$$\sup \Pr\left\{\hat{P} \geq \frac{X_c}{n} \mid P = P_0, n\right\} = \sup \Pr\{X \geq X_c \mid P = P_0, n\} = \sup P_{reject, P_0} = 1 - \beta = 0.95.$$

Thus, as long as

$$\hat{P} \geq \frac{X_c}{n} = P_c$$

we would REJECT the null hypothesis, H_0. With $n = 100$, and $P_0 = 0.95$, $P_c = 92$ would yield $P_{reject,P0} \approx 0.937$. The critical value of 92 gives the closest value of $P_{reject,P0}$ less than or equal to $1 - \beta = 0.95$. In contrast to the quality engineering paradigm, the critical value under the bioavailability paradigm would be to find P_c such that:

$$\sup \Pr\left\{\hat{P} \geq \frac{X_c}{n} \mid P = P_0, n\right\} = \sup \Pr\{X \geq X_c \mid P = P_0, n\} = \sup P_{reject,P0} = \alpha = 0.05.$$

In this case, $X_c = 99$ would yield $P_{reject,P0} = 0.037$, again the closest value less than or equal to 0.05.

Juran and Godfrey (1999) describe an acceptance sampling test for percent defective parts with AQL = 1.2 percent, sample size of $n = 150$, and a critical value of

$$P_c = \frac{X_c}{n} = \frac{4}{150} \approx 2.67\%.$$

In this test, the hypotheses are

$$H_0: P > AQL = 1.2\%$$

$$H_1: P \leq AQL.$$

The critical value of 4 defective parts out of 150 was chosen to give a $100(1 - \beta)$ percent ≈ 96 percent chance of rejecting the null if in fact the percent defect was equal to the AQL = 1.2 percent. The null hypothesis is rejected if the sample percent defective is less than or equal to 2.67 percent, which is greater than the "null" value of AQL = 1.2 percent. So, for example, if 3 defective parts were found in a sample of $n = 150$ (i.e., 2.0 percent), then the null hypothesis would be rejected in favor of the alternative. In the bioavailability paradigm, the critical value would have been 0.0 percent ($\alpha \approx 0.1635$) and therefore with 3 defective parts found the null would not have been rejected.

The notion of noninferiority may bother some people when they realize that the observed value of a sample test statistic can be "worse" than the "null" value, and yet it is possible that they could still reject inferiority in favor of noninferiority. However, consider an analogy to a game of fortune. Suppose your friend told you that you have at least a 90 percent chance of winning at a game. If you played the game 100 times, and won 89 of those times, would you call your friend a liar? After all, 89 is only 89 percent of 100.

Probably you would not suspect your friend of lying. Conversely, if you had only won 50 times, you might at least suspect that your friend was incorrect about the chances of winning. For inferiority hypotheses, we will calculate the worst value we would expect (with probability $1 - \beta = 95$ percent) to observe for a test statistic if the truth is at the border between inferiority and noninferiority. Remember that in the case of the traditional hypothesis test, the observed statistic value can be better than the null parameter value, and we might still fail to reject the null hypothesis.

In fact, the only difference between the bioavailability and quality engineering paradigms is whether or not the boundary conditions for parameters are actually "acceptable" or not. The bioavailability formulation of the equivalence problem gives a small chance (α) of rejecting the null hypothesis when the parameter is at the boundary; the quality engineering paradigm gives a large chance ($1 - \beta$) of rejecting the null hypothesis when the parameter is at the boundary. Bickel and Doksum (2009) give an example of an acceptance sampling test formulated with the bioavailability paradigm. The hypotheses are stated as:

$$H_0: \Pr\{defective\} \geq \theta_0$$

$$H_1: \Pr\{defective\} < \theta_0.$$

In their example, they choose a critical value for the number of defectives, X_c, found in a sample of size n such that:

$$\Pr\{X \leq X_c | \theta_0\} \leq \alpha.$$

That is, if the percent defective is as high as $100\theta_0$ percent, they want to have a small chance (α) of rejecting the null hypothesis. In the paradigm followed in this work, the value θ_0 would the highest *acceptable* level of the parameter, and the test would be chosen to have approximately (up to but not exceeding) a $100(1 - \alpha)$ percent chance of rejecting the null hypothesis.

In either case, whatever critical value is chosen, it is possible to determine parameter values that yield a low (α) or high ($1 - \alpha$) chance of rejecting the null. It is always possible to find conditions (boundaries) so that the critical regions are identical, regardless of whether the boundaries are defined to yield an α or $1 - \alpha$ chance of rejecting the null. That is, if $P1$ is the parameter value that gives an α chance of falling in critical region R, then there is always a parameter value $P2$ that gives (approximately) a $1 - \alpha$ chance that the test statistic will fall in the same region R. So, to elaborate the example of Bickel and Doksum, if they had chosen $\theta_1 = 6.35$ percent as the boundary condition for the hypothesis, then the critical value of 4 defective out of $n = 150$ would yield about $\alpha = 3.5$ percent chance of rejecting the null hypothesis. The same critical value, $4/150 \approx 2.67$ percent, yields about a 96.5 percent $= 100$ percent $- 3.5$ percent chance of rejecting the null if

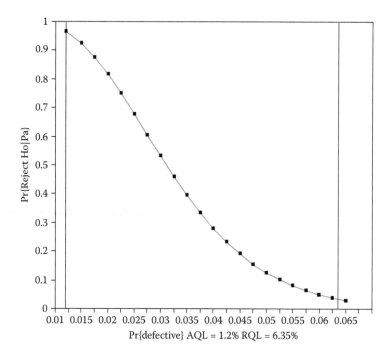

FIGURE 0.1
Power curve for noninferiority test with critical value $4/150 \approx 2.67$ percent.

the parameter $\theta = \theta_2 = 1.2$ percent. In the language of quality engineering (Schilling and Neubauer, 2009), 1.2 percent is the AQL, whereas 6.35 percent would be the rejectable quality limit (RQL). The power curve for the test, regardless of whether the hypotheses are stated with $\theta_1 = 6.35$ percent and tested using the bioavailability paradigm, or with $\theta_2 = 1.2$ percent and tested with the quality engineering paradigm, is identical, and is shown in Figure 0.1. Using the language of quality engineering, this curve could also be referred to as the operating characteristic, or OC curve, in that it describes the probability of passing a quality test. To add some formalism to the reconciliation between the two "paradigms," in the bioavailability paradigm, the hypotheses would be:

$$H_0: \Pr\{defective\} \geq RQL$$

$$H_1: \Pr\{defective\} < RQL$$

and the test, or rejection criterion, would be that the maximum number of defectives, X_c, found in a sample of size n should be such that

$$\Pr\{X \leq X_c | RQL\} \leq \alpha.$$

This test is uniformly most powerful (Bickel and Doksum, 2009). Conversely, in the quality engineering paradigm, the hypotheses would be

$$H_0: \Pr\{defective\} > AQL$$

$$H_1: \Pr\{defective\} \leq AQL$$

and the rejection criterion would be the maximum number of defectives, X_c, found in a sample of size n should be such that

$$\Pr\{X \leq X_c | AQL\} \leq 1 - \beta, \quad \text{with } \beta = \alpha.$$

In other words, AQL and RQL are found such that, if one is fixed, the other can be found to have exactly the same critical region. The difference in the paradigms becomes one of fixing RQL, finding a test with critical region of size α, and then finding an AQL that corresponds to a power of $1 - \beta$, or fixing AQL, finding a test with power $1 - \beta$, and then finding RQL that corresponds to a power of α.

Finally, consider the following example. Suppose you want a probability of "success" to be at least 95 percent. Bickel and Doksum would construct the hypotheses $H_0: P \leq 95$ percent vs. $H_1: P > 95$ percent. They would find a critical value, X_c, for number of successes, X, out of $N = 200$ Bernoulli trials (let's just assume you have a binomial parameter you are trying to make an hypothesis about) such that:

$$\sup \Pr\{X \geq X_c | N, P = 0.95\} = \alpha.$$

The value $X_c = 195$ yields $\alpha = 6.23$ percent, which is the least upper bound (sup, or supremum) for $\Pr\{X \geq X_c | N, P = 0.95\}$ with $N = 200$, $P = 0.95$. That means if $P = 0.95$ exactly, you have only $\alpha = 6.23$ percent chance of rejecting H_0 and concluding noninferiority. But what if $P = 0.95$ is perfectly acceptable? Really, there should be a HIGH chance of rejecting H_0 if $P = 0.95$. So, Bickel and Doksum (and most of the mathematical statisticians) are implying that if $P = 0.95$ really is acceptable, then the hypotheses should be $H_0: P \leq 89.27$ percent vs. $H_1: P > 89.27$ percent.

The critical value, with $N = 200$, is $X_c = 186$, so that $\Pr\{X \geq 186 | N = 200, P = 89.27$ percent$\} = 0.050 = \Pr\{REJECT H_0 | N = 200, P = 89.27$ percent$\}$ and $\Pr\{X > 186 | N = 200, P = 95$ percent$\} = 0.922$. In other words, this critical value of $X_c = 186$ with $N = 200$ yields a high chance of rejecting H_0 if $P = 0.95$, which is what is desired. The test would be constructed in this way in order to fix Type I error risk at α. However, there is no particular reason not to construct a test by fixing β. The probability of Type I error then depends on just how bad the truth is. In other words, we could switch the roles of Type I and Type II errors, so to speak. In this work, we will generally fix β by choosing a critical region such that under the null hypothesis, there is

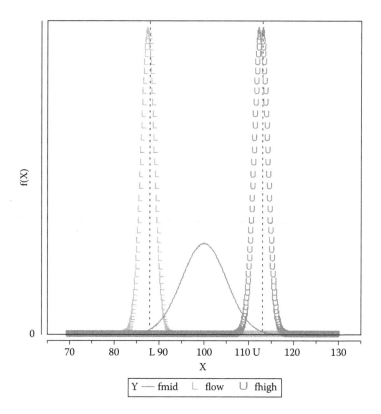

FIGURE 0.2
Illustrating the two equivalence paradigms.

approximately a β chance of FAILing to reject, and then compute a Type I error risk curve.

As a final attempt to illustrate the differences between the two equivalence paradigms, consider Figure 0.2. The values on the horizontal axis represent potential values of the parameter of interest, and L and U represent the lower and upper acceptable limits for that parameter. The frequency curves (f_{mid}, f_{low}, f_{high}) represent the sampling distributions of the test statistic used to decide whether the system(s) under consideration are equivalent, or that the single system under consideration is performing or designed acceptably. In the quality engineering paradigm, L and U would be the most extreme values of the parameter that are completely acceptable. Thus, if the parameter is actually equal to L, we would want the sampling distribution to look something like f_{low}. Or, if the true value of the parameter was equal to U, we would want a sampling distribution to look like f_{high}. The critical values would be the lower 100β percent tail of f_{low}, and the upper 100(1 − β) percent tail of f_{high}. In other words, if the parameter value was equal to L or to U, we would want the critical value to be defined so that there was approximately a $100(1 - \beta)$ percent chance of passing the test. If the bioavailability methods were applied using L and U and the boundaries defined in the hypotheses, then the sample distribution of the test statistic would look more like f_{mid}, and there would be a $100(1 - \beta)$ percent chance of failing the test if the true value of the parameter were equal to either L or U.

A Note on Confidence Regions

In the bioavailability paradigm, the null hypothesis (nonequivalence or inferiority) is rejected if the parameter space for which noninferiority or equivalence is defined completely contains the $1 - \alpha$ confidence region. In the quality engineering paradigm, the null hypothesis is rejected if the "equivalence" or "noninferiority" parameter space has a nonempty intersection with the $1 - \beta$ confidence region.

Organization of the Book

The text is organized into chapters, and each chapter is further divided into tests. Every test will be organized into the following subchapters:

Parameters: a list of the parameters involved, some of which may have known values, and others of which may be the object of the hypothesis test.

Hypotheses: a null (adverse) hypothesis and its logical negation, the alternative, will be stated.

Data: a description of what type of data will be used to make the inferences.

Critical value(s): values of sample statistics that would be decision points, for example, if the statistic exceeds the critical value, then reject the null hypothesis in favor of the alternative.

Discussion: some theoretical and practical considerations related to the hypotheses.

Example: a numerical example.

Confidence interval formulation: formulas or procedures for obtaining confidence regions related to the test.

Computational considerations: some notes about software.

Who Should Use This Book and How to Use It

This book is intended as a handbook for those who are required to demonstrate equivalence or noninferiority, in the quality engineering paradigm, through empirical experimentation. There can be many reasons for such requirements, although process validation and design verification testing are two possible motivations. Engineers and applied scientists are typical of those who have the need to demonstrate equivalence or noninferiority.

If you are an engineer or applied scientist who requires a method for analyzing data from a validation/equivalence/noninferiority test, then:

1. Find the type of test that you are performing in the table of contents.

2. Read the parameters, hypotheses, data, and critical values subsections to determine what inputs you will need and how to decide whether the test was passed or failed.

3. You may want to look at the subsection on computational considerations if you are uncertain as to what functions or procedures of various software packages to use.

4. If you need to compute a power curve for the purposes of documentation for your test plan, or to understand the risk levels of the test, find the theoretical derivations and formulas in the discussion subsection, and look in Chapter 11 on computer code for programs useful for generating power curves.

If you are interested in the basis for the tests, read the discussion subsection for the tests of interest. You may also find the example subsections helpful.

1

Proportions and Binomial Random Variables

Test 1.1 Single Binomial Proportion (One-Sided; Probability of "Success")

<u>Parameters:</u>
P_0 = probability of "success"
n = number of Bernoulli trials

<u>Hypotheses:</u>

$$H_0: P < P_0$$

$$H_1: P \geq P_0$$

<u>Data:</u>

$$X = \text{number of successes out of } n \text{ trials}$$

<u>Critical value(s):</u>

$$X_c = \text{minimum number of successes needed to REJECT } H_0.$$

<u>Discussion:</u>
The experimenter chooses X_c so that:

$$\Pr\{X \geq X_c | P_0\} \approx 1 - \beta \text{ (usually 0.95)}$$

where:

$$\Pr\{X \geq X_c \mid P, n\} = \sum_{k=X_c}^{n} \binom{n}{k} (P)^k (1-P)^{n-k}.$$

The power curve is generated by plotting $\Pr\{X \geq X_c | P_a\}$ on the vertical axis against P_a on the horizontal. For values of $P_a < P_0$, the power will be less than $1 - \beta$.

Example:

It is desired that at least 95 percent of the units of some machine part manufactured satisfy the criteria for some quality inspection. Then $P_0 = 0.95$. To test the hypotheses H_0: $P < 0.95$ versus H_1: $P \geq 0.95$, it was decided to use a sample of $n = 100$ units. Let $1 - \beta = 0.95$. The critical number of "successes" was chose to be $X_c = 92$, since:

$$\sum_{k=92}^{100} \binom{100}{k} (0.95)^k (0.05)^{n-k} \approx 0.9369 \leq 1 - \beta = 0.95.$$

Thus, if $X = 92$ or higher, the null hypothesis would be rejected. Do not be confused by the fact that, in this example, $P_0 = 0.95$ and $1 - \beta = 0.95$. Figure 1.1 shows a power curve for this test.

Confidence interval formulation:

The lower confidence limit, P_L, is the value such that, given X successes observed out of n trials:

$$\sum_{k=X}^{n} \binom{n}{k} P_L^k (1 - P_L)^{n-k} = 1 - \sum_{k=0}^{X-1} \binom{n}{k} P_L^k (1 - P_L)^{n-k} \approx \beta.$$

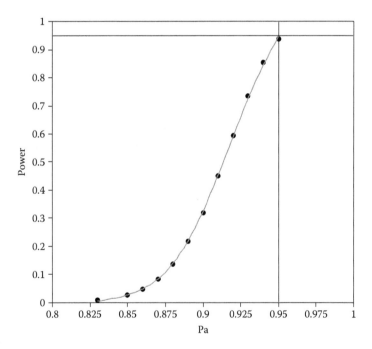

FIGURE 1.1
Power curve for H_0: $P < 0.95$ vs. H_1: $P > = 0.95$.

In the example, if $X = 93$ out of $n = 100$, with $1 - \beta = 0.95$, the value $P_L = 0.8726$ yields:

$$p = \sum_{k=93}^{100} \binom{100}{k} P_L^k (1 - P_L)^{100-k} = 1 - \sum_{k=0}^{92} \binom{100}{k} P_L^k (1 - P_L)^{100-k} \approx 0.0502$$

so $1 - p \approx 0.9498 \leq 1 - \beta = 0.95$, so the lower confidence limit for P is $P_L = 0.8726$.

Computational considerations:

- SAS code

```
libname stuff 'H:\Personal Data\Equivalence & Noninferiority\
Programs & Output';

data calc;
   set stuff.d20121026_test_1_1_example_data;

   xc = 21;
   power = 1 - probbnml(0.8,30,xc-1);/* Pr{Binomial > = xc} */
   run;

proc freq data = calc;
   tables outcome;
   run;

proc print data = calc;/*dataset calc has columns n muL muA
sigma delta nc power */

   run;
```

```
The SAS System          08:19 Friday, October 26, 2012 3

                        The FREQ Procedure

                             outcome

                                    Cumulative Cumulative
           outcome    Frequency   Percent   Frequency   Percent
           ffffffffffffffffffffffffffffffffffffffffffffffff
              0           4        13.33        4        13.33
              1          26        86.67       30       100.00
The SAS System          08:19 Friday, October 26, 2012 4

              Obs       outcome      xc        power
               1           1         21       0.93891
               2           1         21       0.93891
```

3	0	21	0.93891
4	1	21	0.93891
5	1	21	0.93891
6	1	21	0.93891
7	1	21	0.93891
8	1	21	0.93891
9	1	21	0.93891
10	1	21	0.93891
11	1	21	0.93891
12	1	21	0.93891
13	1	21	0.93891
14	1	21	0.93891
15	1	21	0.93891
16	1	21	0.93891
17	1	21	0.93891
18	1	21	0.93891
19	1	21	0.93891
20	1	21	0.93891
21	1	21	0.93891
22	1	21	0.93891
23	1	21	0.93891
24	0	21	0.93891
25	0	21	0.93891
26	1	21	0.93891
27	1	21	0.93891
28	0	21	0.93891
29	1	21	0.93891
30	1	21	0.93891

- JMP data table and formulas:

  Example: $n = 30$; $P_0 = 0.80$, $\beta = 0.05$ (Figure 1.2).

Test 1.2 Single Binomial Proportion (Two-Sided)

Parameters:
  P_0 = probability of "success"
  Δ_0 = allowable difference
  n = number of Bernoulli trials

Hypotheses:

$$H_0: P < P_0 - \Delta_0 \quad \text{OR} \quad P > P_0 + \Delta_0$$

$$H_1: P_0 - \Delta_0 \leq P \leq P_0 + \Delta_0$$

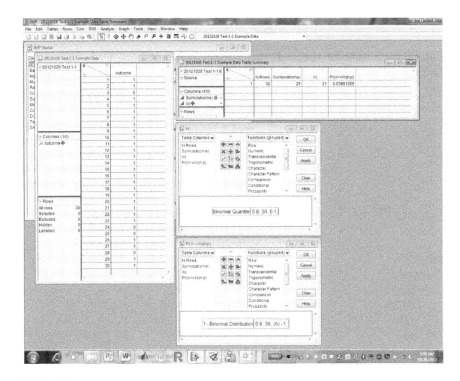

FIGURE 1.2
Test 1.1: JMP screen.

<u>Data:</u>

$$X = \text{number of successes out of } n \text{ trials}$$

<u>Critical value(s):</u>

X_L = minimum number of successes needed to REJECT H_0.

X_U = maximum number of successes needed to REJECT H_0.

If $X < X_L$ or $X > X_U$, then H_0 is rejected in favor of H_1.

<u>Discussion:</u>

As mentioned earlier, in the case of two-sided tests, the two one-sided test (TOST) philosophy will be adopted. Thus, X_L is chosen so that:

$$\Pr\{X \geq X_L \mid P_0 - \Delta_0\} \approx 1 - \beta \text{ (usually 0.95)}$$

and

$$\Pr\{X \le X_U \mid P_0 + \Delta_0\} \approx 1 - \beta \text{ (usually 0.95).}$$

That is, each side of the null hypothesis is treated as a one-sided test. The probability formulas are

$$\Pr\{X \ge X_L \mid P_0 - \Delta_0, n\} = \sum_{k=X_L}^{n} \binom{n}{k} (P_0 - \Delta_0)^k (1 - P_0 + \Delta_0)^{n-k}$$

and

$$\Pr\{X \le X_U \mid P_0 + \Delta_0, n\} = \sum_{k=0}^{X_U} \binom{n}{k} (P_0 + \Delta_0)^k (1 - P_0 - \Delta_0)^{n-k}.$$

Example:

  Suppose

  $P_0 = 0.50$

  $\Delta_0 = 0.025$

  $n = 100$

  $X = 59$

  $1 - \beta = 0.95$

Then

$$P_L = P_0 - \Delta_0 = 0.50 - 0.025 = 0.475$$

$$P_U = P_0 + \Delta_0 = 0.50 + 0.025 = 0.525$$

X_L is the value such that

$$\Pr\{X \ge X_L \mid P_0 - \Delta_0\} = \sum_{k=X_L}^{n} \binom{n}{k} (0.475)^k (1 - 0.475)^{n-k} \approx 0.95$$

and X_U is the value such that

$$\Pr\{X \le X_U \mid P_0 + \Delta_0\} = \sum_{k=0}^{X_U} \binom{n}{k} (0.525)^k (1 - 0.525)^{n-k} \approx 0.95.$$

So $X_L = 40$, and

$$\sum_{k=X_L}^{n} \binom{n}{k} (0.475)^k (1 - 0.475)^{n-k} \approx 0.9460$$

and $X_U = 60$, so

$$\sum_{k=0}^{X_U} \binom{n}{k} (0.525)^k (1 - 0.525)^{n-k} \approx 0.9460.$$

Since $40 < 59 < 60$, the null hypothesis is rejected.

Confidence interval formulation:
   Let

$$\hat{P} = \frac{X}{n}.$$

Then P_L is the value such that

$$\sum_{k=X}^{n} \binom{n}{k} (P_L)^k (1 - P_L)^{n-k} \approx \beta.$$

and P_U is the value such that

$$\sum_{k=0}^{X} \binom{n}{k} (P_U)^k (1 - P_U)^{n-k} \approx \beta.$$

   Then (P_L, P_U) is a $100(1 - 2\beta)$ percent confidence interval for P. In the example, with $n = 100$, $X = 59$, $P_L \approx 0.503$, and $P_U \approx 0.673$.

Computational considerations:
   The same procedures used for Test 1.1 apply to Test 1.2.

Test 1.3 Difference of Two Proportions (Two-Sided)

Parameters:
   $p_1 = \Pr\{\text{"success," group or treatment 1}\}$
   $p_2 = \Pr\{\text{"success," group or treatment 2}\}$
   $n_1 = $ sample size, group or treatment 1
   $n_2 = $ sample size, group or treatment 2
   $\Delta_L = $ minimum allowable difference between p_1 and p_2
   $\Delta_H = $ maximum allowable difference between p_1 and p_2

Hypotheses:

$$H_0: p_1 - p_2 < \Delta_L \quad OR \quad p_1 - p_2 > \Delta_H$$

$$H_1: \Delta_L \leq p_1 - p_2 \leq \Delta_H$$

Data:

$$X_1 = \text{number of "success," group or treatment 1}$$

$$X_2 = \text{number of "success," group or treatment 2}$$

$$\hat{p}_1 = \frac{X_1}{n_1}$$

$$\hat{p}_2 = \frac{X_2}{n_2}$$

Critical value(s):
Reject H_0 if

$$\hat{p}_1 - \hat{p}_2 + z_{1-\beta}SE \geq \Delta_L \quad \text{and} \quad \hat{p}_1 - \hat{p}_2 - z_{1-\beta}SE \leq \Delta_U, \text{ where:}$$

$$SE = \sqrt{\frac{\hat{p}_1(1-\hat{p}_1)}{n_1} + \frac{\hat{p}_2(1-\hat{p}_2)}{n_2}}$$

$z_{1-\beta}$ = the upper $100(1 - \beta)$ percentile of a standard normal distribution.

Discussion:
The exact distribution of the difference between two proportions was derived by Nadarajah and Kotz (2007). However, finding percentiles of this distribution would be at best a difficult numerical approximation exercise. Therefore, only the normal approximation formulas and methodology are presented. The method provides the most accurate results when the population proportions in question are "close" to 0.50 (Armitage, 1971).

Using the normal approximation, the power to reject the null hypothesis is given by

$$\Pr\left\{ z \geq \frac{\Delta_L - \Delta}{SE} - z_{1-\beta} \right\} \text{ or}$$

$$\Pr\left\{ z \leq \frac{\Delta_U - \Delta}{SE} + z_{1-\beta} \right\} \text{ where } z \sim N(0, 1).$$

Example:
  Suppose
  $n_1 = n_2 = 100$
  $X_1 = 50$
  $X_2 = 62$
  $\Delta_L = -0.01$
  $\Delta_H = +0.01$
Then

$$\hat{p}_1 = \frac{X_1}{n_1} = \frac{50}{100} = 0.50$$

$$\hat{p}_2 = \frac{X_2}{n_2} = \frac{62}{100} = 0.62$$

$$SE = \sqrt{\frac{\hat{p}_1(1-\hat{p}_1)}{n_1} + \frac{\hat{p}_2(1-\hat{p}_2)}{n_2}} = \sqrt{\frac{0.50(0.50)}{100} + \frac{0.62(0.38)}{100}} \approx 0.0697$$

$$1 - \beta = 0.95$$

$$z_{1-\beta} \approx 1.645.$$

The critical values are

$$p_1 - p_2 + z_{1-\beta}SE \approx -0.0054 \geq -0.01 = \Delta_L^{*}$$

and

$$p_1 - p_2 - z_{1-\beta}SE \approx -0.2346 \leq +0.01 = \Delta_U.$$

Therefore, the null hypothesis of nonequivalence is rejected. Figure 1.3 illustrates the power curve for this example. Note that only the curve for Δ_L is presented.
Confidence interval formulation:
  The interval

$$\left(\hat{p}_1 - \hat{p}_2 - z_{1-\beta}SE, \hat{p}_1 - \hat{p}_2 + z_{1-\beta}SE \right)$$

is a $100(1 - 2\beta)$ percent confidence interval for $p_1 - p_2$.

---

* Wellek (2003) has suggested to use odd's ratio in lieu of differences between two proportions. However, no text for equivalence based on odd's ratios will be presented in this book.

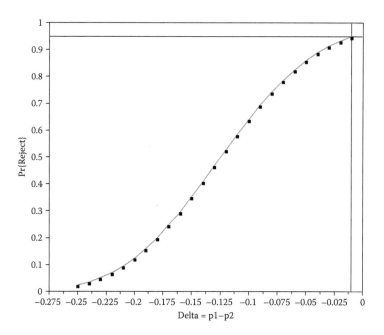

FIGURE 1.3
Power curve for $p_1 - p_2 \leq \Delta_L = -0.01$.

Computational considerations:
   Example: $n = 30$, $\beta = 0.05$, $\Delta_L = 0.05$, $\Delta_U = 0.05$

   • SAS code

```
libname stuff 'H:\Personal Data\Equivalence & Noninferiority\
Programs & Output';

data calc;
   set stuff.d20121026_test_1_3_example_data;

   delL = -0.05;
   delU = 0.05;

   run;

proc freq data = calc;
   tables group1/out = prop1;
   tables group2/out = prop2;
   run;

data prop_first;
   set prop1;
   count1 = COUNT;
```

```
  p1 = PERCENT/100;
  run;
data prop_second;
  set prop2;
  count2 = COUNT;
  p2 = PERCENT/100;
  run;

data prop_all;
  set prop_first;
  set prop_second;
  run;

data comp;
  set prop_all;
  if (group1 = 1 & group2 = 1);
  n1 = count1/p1;
  n2 = count2/p2;
  se = sqrt((p1*(1-p1)/n1)+(p2*(1-p2)/n2));
  low = p1 - p2 + probit(.95)*se;
  high = p1 - p2 - probit(.95)*se;

proc print data = calc;/*dataset calc group1 group2 delL delU
SE High Low*/

  run;

proc print data = comp;

  run;
```

The SAS System   08:19 Friday, October 26, 2012 26

The FREQ Procedure

group1

| | | | Cumulative | Cumulative |
| group1 | Frequency | Percent | Frequency | Percent |
| ffffffffffffffffffffffffffffffffffffffffffffffffffffffffffffffff |
| 0 | 4 | 13.33 | 4 | 13.33 |
| 1 | 26 | 86.67 | 30 | 100.00 |

group2

| | | | Cumulative | Cumulative |
| group2 | Frequency | Percent | Frequency | Percent |
| ffffffffffffffffffffffffffffffffffffffffffffffffffffffffffffffff |
| 0 | 6 | 20.00 | 6 | 20.00 |
| 1 | 24 | 80.00 | 30 | 100.00 |

The SAS System   08:19 Friday, October 26, 2012 27

Obs	group1	group2	delL	delU
1	1	0	-0.05	0.05
2	1	1	-0.05	0.05
3	0	1	-0.05	0.05
4	1	0	-0.05	0.05
5	1	1	-0.05	0.05
6	1	1	-0.05	0.05
7	1	1	-0.05	0.05
8	1	0	-0.05	0.05
9	1	1	-0.05	0.05
10	1	1	-0.05	0.05
11	1	1	-0.05	0.05
12	1	1	-0.05	0.05
13	1	1	-0.05	0.05
14	1	1	-0.05	0.05
15	1	1	-0.05	0.05
16	1	1	-0.05	0.05
17	1	1	-0.05	0.05
18	1	1	-0.05	0.05
19	1	1	-0.05	0.05
20	1	0	-0.05	0.05
21	1	0	-0.05	0.05
22	1	1	-0.05	0.05
23	1	0	-0.05	0.05
24	0	1	-0.05	0.05
25	0	1	-0.05	0.05
26	1	1	-0.05	0.05
27	1	1	-0.05	0.05
28	0	1	-0.05	0.05
29	1	1	-0.05	0.05
30	1	1	-0.05	0.05

The SAS System   08:19 Friday, October 26, 2012 28

Obs	group1	COUNT	PERCENT	count1	p1	group2	count2	p2	n1	n2	se	low	high
1	1	24	80	26	0.86667	1	24	0.8	30	30	0.095839	0.22431	-0.090975

- SAS code (alternative)

```
libname stuff 'H:\Personal Data\Equivalence & Noninferiority\
Programs & Output';

data calc;

  set stuff.d20121026_test_1_3_example_data;

  run;
```

```
proc means data = calc;
  var group1 group2;
  output out = props MEAN = prop1 prop2 N = n1 n2;
  run;

data outcalc;

  set props;

  delL = -0.05;
  delU = 0.05;
  se = sqrt((prop1*(1-prop1)/n1)+(prop2*(1-prop2)/n2));
  low = prop1 - prop2 + probit(.95)*se;
  high = prop1 - prop2 - probit(.95)*se;
  run;

proc print data = outcalc;/* has vars prop1 prop2 n1 n2 delL
delU se low high */

  run;
```

```
        The SAS System        07:03 Sunday, October 28, 2012 11

Obs _TYPE_   _FREQ_   prop1   prop2  n1  n2  delL  delU    se      low      high

 1     0       30     0.86667  0.8   30  30  -0.05  0.05  0.095839  0.22431 -0.090975
```

- JMP Data Table and formulas: see Figure 1.4.

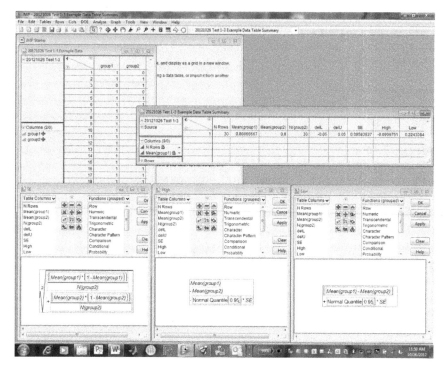

FIGURE 1.4
Test 1.3: JMP screen.

2

Means

Test 2.1 Single Mean (One-Sided)

Parameters:
 μ = population mean
 μ_0 = minimum desired value for μ
 σ = population standard deviation

Hypotheses:

$$H_0: \mu < \mu_0$$

$$H_1: \mu \geq \mu_0$$

Data:

$$\bar{X} = \text{sample mean}$$

$$S = \text{sample standard deviation}$$

$$n = \text{sample size}$$

Critical value(s):
 Reject H_0 if:

$$\bar{X} + t_{1-\beta}\frac{S}{\sqrt{n}} \geq \mu_0$$

where $t_{1-\beta} = 100*(1-\beta)$ percentile of a central t distribution with $n-1$ degrees of freedom.

Discussion:

If $\mu = \mu_0$, then we would expect the sample mean, $\bar{X}$, to be greater than:

$$\mu_0 - t_{1-\beta} \frac{S}{\sqrt{n}}$$

with probability $1 - \beta$. If

$$\bar{X} \geq \mu_0 - t_{1-\beta} \frac{S}{\sqrt{n}}$$

we would reject the hypothesis that $\mu < \mu_0$. Or, in other words, we reject H_0 if

$$\bar{X} + t_{1-\beta} \frac{S}{\sqrt{n}} \geq \mu_0.$$

If the inequalities in the hypotheses are reversed, that is,

$$H_0: \mu > \mu_0$$

$$H_1: \mu \leq \mu_0$$

then we would reject H_0 if

$$\bar{X} - t_{1-\beta} \frac{S}{\sqrt{n}} \leq \mu_0.$$

The power for the former case is given by:

$$\Pr\left\{\bar{X} + t_{1-\beta} \frac{S}{\sqrt{n}} \geq \mu_0 \,\middle|\, \mu = \mu_a, n\right\} = \Pr\left\{\frac{\sqrt{n}\left(\bar{X} - \mu_0\right)}{S} \geq -t_{1-\beta} \,\middle|\, n, \delta = \frac{\sqrt{n}\left(\mu_a - \mu_0\right)}{\sigma}\right\}$$

where

$$\delta = \frac{\sqrt{n}\left(\mu_a - \mu_0\right)}{\sigma}$$

is the noncentrality parameter for the statistic

$$\frac{\bar{X} - \mu_0}{S/\sqrt{n}}$$

which has a noncentral t distribution with $n-1$ degrees of freedom (Johnson, Kotz, and Balakrishnan, 1995).

Example:
   Suppose we hypothesize that the average time until failure of a machine is at least 100 hours. The hypotheses are

$$H_0: \mu < 100$$

versus the alternative

$$H_1: \mu \geq 100.$$

We obtain, from a sample of $n = 20$ times to failure, the data

$$\bar{X} = 99.9 \text{ hours}$$

$$S = 3.4 \text{ hours.}$$

We choose $1 - \beta = 0.95$, so $t_{1-\beta} \approx 1.729$. Thus

$$\bar{X} + t_{1-\beta} \frac{S}{\sqrt{n}} = 99.9 + 1.729 \frac{3.4}{\sqrt{20}} \approx 101.2 > 100.$$

Therefore, we reject the null hypothesis in favor of the alternative.
   To calculate the power under alternative values of $\mu = \mu_a$, we must calculate the probability

$$\Pr\left\{ \bar{X} + t_{1-\beta} \frac{S}{\sqrt{n}} \geq \mu_0 \,|\, \mu = \mu_a, n \right\}.$$

If $\mu = \mu_0$, then

$$t = \frac{\bar{X} - \mu_0}{S/\sqrt{n}}$$

has a central t-distribution with $n-1$ degrees of freedom, and

$$\Pr\left\{ \bar{X} + t_{1-\beta} \frac{S}{\sqrt{n}} \geq \mu_0 \,|\, \mu = \mu_0, n \right\} = \Pr\left\{ \frac{\sqrt{n}(\bar{X} - \mu_0)}{S} \geq -t_{1-\beta} \,|\, \mu = \mu_0, n \right\} = 1 - \beta.$$

If $\mu = \mu_a < \mu_0$, then

$$\frac{\bar{X} - \mu_0}{S/\sqrt{n}}$$

has a noncentral *t*-distribution with noncentrality parameter

$$\delta = \frac{\sqrt{n}\left(\mu_a - \mu_0\right)}{\sigma}.$$

Since, in general, σ is unknown, the power may be plotted against

$$\frac{\delta}{\sqrt{n}} = \frac{\left(\mu_a - \mu_0\right)}{\sigma}$$

or in other words, the difference $\mu_a - \mu_0$ expressed as a proportion of the standard deviation of the population. Note that if $\mu_a = \mu_0$, then $\delta = 0$. Figure 2.1 shows a power curve for the example.

<u>Confidence interval formulation:</u>
    The expression:

$$\bar{X} + t_{1-\beta}\frac{S}{\sqrt{n}}$$

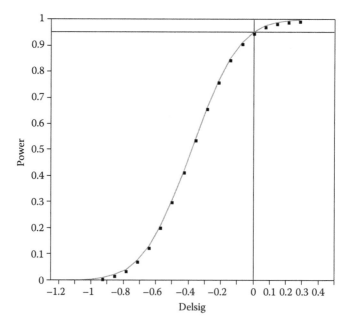

FIGURE 2.1
Test 2.1, power curve *n* = 20.

is a one-sided upper confidence limit for μ. Thus, from the example, the upper 95 percent confidence limit for μ is

$$\bar{X} + t_{1-\beta} \frac{S}{\sqrt{n}} = 99.9 + 1.729 \frac{3.4}{\sqrt{20}} \approx 101.2.$$

Computational considerations:
  Example: $n = 25$, $\beta = 0.05$, $\mu_0 = 10.00$

- SAS code

```
libname stuff 'H:\Personal Data\Equivalence & Noninferiority\
Programs & Output';

data calc;

  set stuff.d20121029_test_2_1_example_data;

  run;

proc means data = calc;
  var X mu0 beta;
  output out = onemean MEAN = xbar popmu betaprob STD = sd N =
  n1;
  run;

data outcalc;

  set onemean;
  se = sd/sqrt(n1);
  lowlim = xbar + tinv(1-betaprob,n1-1)*se;
  run;

proc print data = outcalc;/* has vars xbar popmu betaprob n1
se lowlim */

  run;
```

```
        The SAS System        07:03 Sunday, October 28, 2012 13

Obs _TYPE_ _FREQ_ xbar popmu betaprob sd   n1    se     lowlim
 1     0     25   9.68 10.00  0.05   1.00  25 0.20092 10.0266
```

- JMP Data Table and formulas (Figure 2.2):

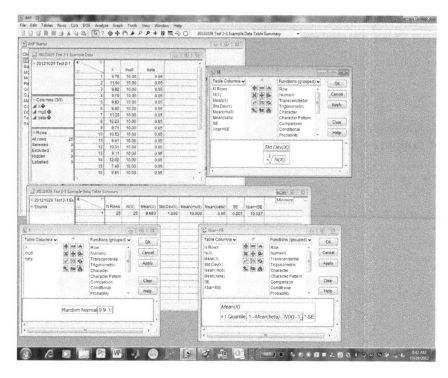

FIGURE 2.2
Test 2.1, JMP screen.

Test 2.2 Comparison of Two Means—Two Independent Samples, Fixed Δ Paradigm (One-Sided)

<u>Parameters:</u>

μ_1 = population mean, "group" 1
μ_2 = population mean, "group" 2
σ_1 = population standard deviation, "group" 1
σ_2 = population standard deviation, "group" 2
Δ_0 = maximum allowable difference between μ_1 and μ_2

<u>Hypotheses:</u>

$$H_0: \mu_1 < \mu_2 - \Delta_0$$

$$H_1: \mu_1 \geq \mu_2 - \Delta_0$$

<u>Data:</u>

$$\bar{X}_1 = \text{sample mean, "group" 1}$$

S_1 = sample standard deviation, "group" 1

n_1 = sample size, "group" 1

$\bar{X}_2$ = sample mean, "group" 2

S_2 = sample standard deviation, "group" 2

n_2 = sample size, "group" 2

Critical value(s):
  Reject H_0 if:

$$\bar{X}_1 - \bar{X}_2 + t_{1-\beta}SE \geq -\Delta_0$$

where $t_{1-\beta}$ = 100*(1 − β) percentile of a central *t*-distribution with $n_1 + n_2 - 2$ degrees of freedom and SE is the standard error for the difference of two means:

$$SE = \sqrt{\frac{S_1^2}{n_1} + \frac{S_2^2}{n_2}} .$$

Discussion:
  If $\mu_1 = \mu_2 - \Delta_0$ exactly, then we would expect that $\bar{X}_1 > \bar{X}_2 - \Delta_0$ about as frequently as $\bar{X}_1 < \bar{X}_2 - \Delta_0$. Since $\mu_1 = \mu_2 - \Delta_0$ would be minimally acceptable, we would want to avoid failing to conclude that $\mu_1 \geq \mu_2 - \Delta_0$ just because $\bar{X}_1 < \bar{X}_2 - \Delta_0$. That is, we would want to conclude that $\mu_1 < \mu_2 - \Delta_0$ only when $\bar{X}_1$ was sufficiently less than $\bar{X}_2$. In other words, we are willing to believe that $\mu_1 \geq \mu_2 - \Delta_0$ (i.e., the alternate hypothesis) as long as

$$\bar{X}_1 \geq \bar{X}_2 - \Delta_0 - t_{1-\beta}SE.$$

As in the case of the single mean, the test statistic under various alternate hypotheses has a noncentral *t*-distribution with $n_1 + n_2 - 2$ degrees of freedom and noncentrality:

$$\delta = \frac{\Delta_a - \Delta_0}{\sqrt{\sigma_1^2/n_1 + \sigma_2^2/n_2}} .$$

Welch (1947) provided an alternative calculation for the degrees of freedom of the two-sample *t*-test, when it is assumed that the variances for the two populations or systems are not equal.

Let:

$$W_1 = \frac{\left(\dfrac{S_1^2/n_1}{S_1^2/n_1 + S_2^2/n_2}\right)^2}{(n_1 - 1)}$$

and

$$W_2 = \frac{\left(\dfrac{S_2^2/n_2}{S_1^2/n_1 + S_2^2/n_2}\right)^2}{(n_2 - 1)}.$$

Then the degrees of freedom for Welch's *t*-test are

$$d.f. = \frac{1}{W_1 + W_2}$$

For simplicity, the conventional $n_1 + n_2 - 2$ degrees of freedom will be used for the examples presented. In actual practice, Welch's formula is recommended.

Power calculations are made as a function of the noncentrality parameter, and particularly as a function of Δ_a.

Example:

Suppose we hypothesize that the mean of "group" 1 is no more than $\Delta_0 = 5.0$ units less than the mean of "group" 2. The data are

$$\bar{X}_1 = 96.0$$
$$S_1 = 2.40$$
$$n_1 = 12$$
$$\bar{X}_2 = 101.5$$
$$S_2 = 2.20$$
$$n_2 = 14$$

We choose $1 - \beta = 0.95$, so $t_{1-\beta} = 1.711$ ($12 + 14 - 2 = 24$ degrees of freedom). We compute

$$SE = \sqrt{\frac{S_1^2}{n_1} + \frac{S_2^2}{n_2}} = \sqrt{\frac{2.40^2}{12} + \frac{2.20^2}{14}} \approx 0.909.$$

The critical value is

$$\bar{X}_1 - \bar{X}_2 + t_{1-\beta}SE = 96 - 101.5 + 1.711 * 0.909 \approx -3.94 \geq -5.0.$$

Therefore, we reject the null hypothesis, H_0 in favor of the alternate, H_1.
If $\mu_1 = \mu_2 - 5.0$, then

$$t = \frac{\bar{X}_1 - \bar{X}_2 + 5.0}{\sqrt{\dfrac{S_1^2}{n_1} + \dfrac{S_2^2}{n_2}}}$$

has a central t-distribution with $n_1 + n_2 - 2$ degrees of freedom.
Thus, the probability of rejecting the null hypothesis when $\mu_1 = \mu_2 - 5.0$ is

$$\Pr\left\{\bar{X}_1 - \bar{X}_2 + t_{1-\beta}SE \geq -5.0\right\} = 1 - \beta.$$

Under some specific alternate hypothesis, such as $\mu_1 = \mu_2 - \Delta_a$, where $\Delta_a > 5.0$, then

$$t = \frac{\bar{X}_1 - \bar{X}_2 + 5.0}{\sqrt{\dfrac{S_1^2}{n_1} + \dfrac{S_2^2}{n_2}}}$$

has a noncentral t-distribution with $n_1 + n_2 - 2$ degrees of freedom and non-centrality parameter

$$\delta = \frac{\Delta_a - 5.0}{\sqrt{\sigma_1^2/n_1 + \sigma_2^2/n_2}}.$$

To calculate a power curve, we will make the simplifying assumptions that

$$\sigma_1 = \sigma_2 = \sigma \quad \text{and} \quad n_1 = n_2 = \frac{n_1 + n_2}{2} = 13.$$

Thus, the noncentrality parameter simplifies to

$$\delta = \frac{\sqrt{n}\,(\Delta - 5.0)}{\sqrt{2}\sigma}.$$

Expressing

$$\frac{\Delta_a - 5.0}{\sigma}$$

as a proportion (i.e., the difference in σ units) is usually easier than obtaining a reasonable estimate of σ. Thus, the power curve will be expressed as a function of

$$\gamma = \frac{\Delta_a - 5.0}{\sigma}$$

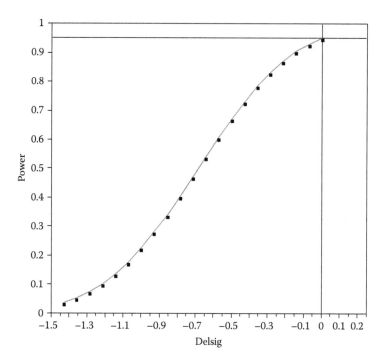

FIGURE 2.3
Test 2.2, power curve for equivalence of two means.

for $\gamma = 0, \ldots, 2.0$ (σ units). Figure 2.3 shows the power curve for this test.

Confidence interval formulation:
  The expression

$$\overline{X}_1 - \overline{X}_2 + t_{1-\beta}SE$$

is a one-sided $100(1 - \beta)$ percent upper confidence limit for $\mu_1 - \mu_2$. From the example, the upper 95 percent confidence limit for $\mu_1 - \mu_2$ is

$$\overline{X}_1 - \overline{X}_2 + t_{1-\beta}SE = 96 - 101.5 + 1.711 * 0.909 \approx -3.94.$$

Computational considerations:
  For this test, one could use a confidence limit computed by various programs for the difference between two means. The upper confidence limit should be compared to the lower "acceptable" bound on the difference, and a one-sided $100(1 - \beta)$ percent [or a two-sided $100(1 - 2\beta)$ percent] limit should be computed, in concert with the two one-sided test (TOST) philosophy.

Some software might as a default compute two-sided 95 percent confidence limits, which, if used to accept or reject H_0, would be less stringent than the two one-sided limits.

- SAS code

```
libname stuff 'H:\Personal Data\Equivalence & Noninferiority\
Programs & Output';

data calc;

  set stuff.d20121102_test_2_2_fixed;

  run;

proc means data = calc;
  var X1 X2 del0 beta;
  output out = onemean MEAN = xbar1 xbar2 pdel0 betaprob STD =
  sd1 sd2 N = n1 n2;
  run;

data outcalc;

  set onemean;
  se = sqrt(sd1**2/n1 + sd2**2/n2);
  w1 = ((sd1**2/n1)/(sd1**2/n1 + sd2**2/n2))**2/(n1-1);
  w2 = ((sd2**2/n1)/(sd1**2/n1 + sd2**2/n2))**2/(n2-1);
  dfe = 1/(w1 + w2);
  lowlim = xbar1 - xbar2 + tinv(1-betaprob,dfe)*se;
  run;

proc print data = outcalc;/* has vars xbar1 xbar2 pdel0
betaprob n1 n2 se lowlim */

  run;
```

```
          The SAS System          10:16 Friday, November 2, 2012 10

Obs   _TYPE_    _FREQ_    xbar1    xbar2    pdel0    betaprob  sd1

1        0        25       9.68    11.42     1.10      0.05    1.00

Obs  sd2   n1   n2     se       w1           w2        dfe      lowlim

1   1.03   25   25  0.28816 .009847617  0.011002   47.9632  -1.25322
```

- JMP Data Table and formulas (shown in Figures 2.4, 2.5, and 2.6)

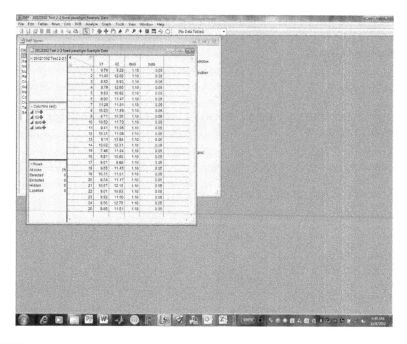

FIGURE 2.4
Test 2.2, JMP screen 1.

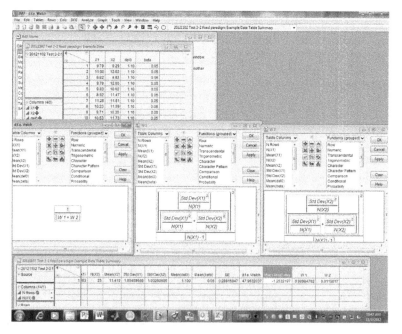

FIGURE 2.5
Test 2.2, JMP screen 2.

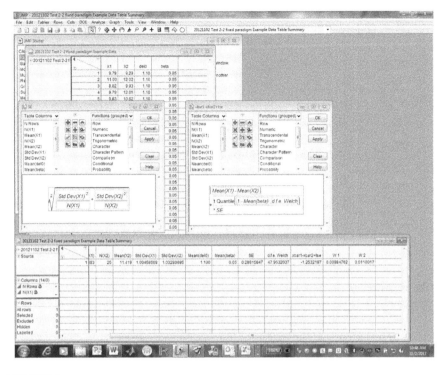

FIGURE 2.6
Test 2.2, JMP screen 3.

Test 2.3 Comparison of Two Means—Proportional Difference Paradigm for Two Independent Samples (One-Sided)

Parameters:
μ_1 = population mean, "group" 1
μ_2 = population mean, "group" 2
σ_1 = population standard deviation, "group" 1
σ_2 = population standard deviation, "group" 2
$1 - p_0$ = minimum proportion of μ_2 that equals μ_1
In other words, we require

$$\frac{\mu_1}{\mu_2} \geq 1 - p_0 \Rightarrow \mu_1 \geq (1 - p_0)\mu_2.$$

The two paradigms for comparing means are related, that is, $\mu_2 - \Delta_0 = (1 - p_0)$ μ_2. So $1 - p_0$ is the desired minimum value of some proportion $1 - p$.

Hypotheses:

$$H_0: \mu_1 < (1 - p_0)\mu_2$$

$$H_1: \mu_1 \geq (1 - p_0)\mu_2$$

Data:

$\bar{X}_1$ = sample mean, "group" 1

S_1 = sample standard deviation, "group" 1

n_1 = sample size, "group" 1

$\bar{X}_2$ = sample mean, "group" 2

S_2 = sample standard deviation, "group" 2

n_2 = sample size, "group" 2

Critical value(s):
  Reject H_0 if:

$$\bar{X}_1 - (1 - p_0)\bar{X}_2 + t_{1-\beta}SE \geq 0$$

where $t_{1-\beta}$ = 100*(1 − β) percentile of a central *t*-distribution with $n_1 + n_2 - 2$ degrees of freedom and SE is the standard error for the difference $\bar{X}_1 - (1 - p_0)\bar{X}_2$.

$$SE = \sqrt{\frac{S_1^2}{n_1} + \frac{(1-p_0)^2 S_2^2}{n_2}}.$$

Discussion:
  If $\mu_1 = (1 - p_0)\mu_2$ exactly, then we would expect that $\bar{X}_1 > (1-p_0)\bar{X}_2$ about as frequently as $\bar{X}_1 < (1-p_0)\bar{X}_2$. Since $\mu_1 = (1 - p_0)\mu_2$ would be minimally acceptable, we would want to avoid failing to conclude that $\mu_1 \geq (1 - p_0)\mu_2$ just because $\bar{X}_1 < (1-p_0)\bar{X}_2$. That is, we would want to conclude that $\mu 1 < (1 - p0)\mu_2$ only when $\bar{X}_1$ was sufficiently less than $(1-p_0)\bar{X}_2$. In other words, we are willing to believe that $\mu_1 \geq (1 - p_0)\mu_2$ (i.e., the alternate hypothesis) as long as $\bar{X}_1 \geq (1-p_0)\bar{X}_2 - t_{1-\beta}SE$.
  One might be tempted to estimate the parameter

$$\frac{\mu_1}{\mu_2} = 1 - p$$

with the sample statistic

$$\frac{\bar{X}_1}{\bar{X}_2} = 1 - \hat{p}.$$

There are several problems with such an estimator. Even if it was true that

$$\frac{\mu_1}{\mu_2} = 1 - p_0,$$

the sampling variation in the ratio statistic

$$\frac{\overline{X}_1}{\overline{X}_2}$$

is intractable mathematically (Springer, 1979). Thus, it is not possible to know in any sense how close to the truth the sample ratio is. It might be possible to use a resampling technique such as the bootstrap (Efron, 1990) to assess a confidence interval for the parameter $1 - p$.

The power to reject the null hypothesis is based on a noncentral t-distribution with $n_1 + n_2 - 2$ degrees of freedom and noncentrality parameter

$$\delta = \frac{(p_0 - p_a)\mu_2 \sqrt{n}}{\sigma \sqrt{1 + (1 - p_0)^2}}$$

where p_a is the alternate hypothesis value for the proportion, p.

Example:
  Let $p_0 = 0.05$ ($1 - p_0 = 0.95$). The data are

  $\overline{X}_1 = 93.5$

  $S_1 = 2.5$

  $\overline{X}_2 = 100.0$

  $S_2 = 3.1$

  $n_1 = n_2 = 15$

so

$$SE = \sqrt{\frac{S_1^2}{n_1} + \frac{(1 - p_0)^2 S_2^2}{n_2}} = \sqrt{\frac{2.5^2}{15} + \frac{(0.95)^2 3.1^2}{15}} \approx 0.997.$$

Let $1 - \beta = 0.95$. Then with $15 + 15 - 2 = 28$ degrees of freedom, $t_{1-\beta} = 1.701$ and $\overline{X}_1 - (1 - p_0)\overline{X}_2 + t_{1-\beta}SE = 93.5 - (0.95)*100.0 + (1.701)*0.997 \approx 0.197 > 0$. Therefore, the data lead us to reject the null in favor of the alternate hypothesis that $\mu_1 \geq (0.95)\mu_2$. Note that had $\overline{X}_1 = 93.3$, and all the other data had remained the same, we would have failed to reject the null hypothesis.

Power calculations are similar to those for the Δ paradigm, with some important differences.

Note that under the condition that $\mu_1 = (1 - p_a)\mu_2, \quad p_a \neq p_0$

$$E\left[\bar{X}_1 - (1-p_0)\bar{X}_2\right] = (1-p_a)\mu_2 - (1-p_0)\mu_2 = (p_0 - p_a)\mu_2$$

As a result, the noncentrality parameter for the test statistic is

$$\frac{(p_0 - p_a)\mu_2\sqrt{n}}{\sigma\sqrt{1+(1-p_0)^2}}$$

Thus, in order to compute the probability of rejecting the null, estimates of both σ and μ_2 are required. Figure 2.7 shows the power curve for the example, assuming $\mu_2 = 100$ and $\sigma = 2.8$. In this example, Pr{Reject $H_0|p_a = 0.0835\} \approx 0.051$.

Confidence interval formulation:
   The expression

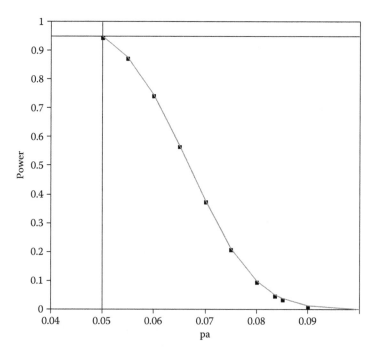

FIGURE 2.7
Test 2.3, power curve for equivalence of two means—percent paradigm.

$$\bar{X}_1 - (1-p_0)\bar{X}_2 + t_{1-\beta}SE$$

is a one-sided $100(1-\beta)$ percent upper confidence limit for $\mu_1 - (1-p_0)\mu_2$. From the example, the 95 percent upper confidence limit for $\mu_1 - (1-p_0)\mu_2$ is

$$\bar{X}_1 - (1-p_0)\bar{X}_2 + t_{1-\beta}SE = 93.5 - (0.95)*100.0 + (1.701)*0.997 \approx 0.197.$$

<u>Computational considerations:</u>

* SAS code

```
libname stuff 'H:\Personal Data\Equivalence & Noninferiority\
Programs & Output';

data calc;

  set stuff.d20121104_test_2_3_prop;

  run;

proc means data = calc;
  var X1 X2 p0 beta;
  output out = onemean MEAN = xbar1 xbar2 p0val betaprob STD =
  sd1 sd2 N = n1 n2;
  run;

data outcalc;

  set onemean;
  se = sqrt(sd1**2/n1 + ((1-p0val)**2)*sd2**2/n2);
  w1 = ((sd1**2/n1)/(sd1**2/n1 + sd2**2/n2))**2/(n1-1);
  w2 = ((sd2**2/n1)/(sd1**2/n1 + sd2**2/n2))**2/(n2-1);
  dfe = 1/(w1 + w2);
  lowlim = xbar1 - (1-p0val)*xbar2 + tinv(1-betaprob,dfe)*se;
  run;

proc print data = outcalc;/* has vars xbar1 xbar2 p0val
betaprob n1 n2 se lowlim */

  run;
```

```
                    The MEANS Procedure

Variable  Label   N      Mean    Std Dev    Minimum     Maximum
ƒƒƒƒƒƒƒƒƒƒƒƒƒƒƒƒƒƒƒƒƒƒƒƒƒƒƒƒƒƒƒƒƒƒƒƒƒƒƒƒƒƒƒƒƒƒƒƒƒƒƒƒƒƒƒƒƒƒƒƒƒƒ
X1        X1     25  8.8799300  1.0305582  7.3220509  11.6006410
X2        X2     25  9.6829027  1.0045857  7.4892100  12.0170228
p0        p0     25  0.1000000          0  0.1000000   0.1000000
beta      beta   25  0.0500000          0  0.0500000   0.0500000
ƒƒƒƒƒƒƒƒƒƒƒƒƒƒƒƒƒƒƒƒƒƒƒƒƒƒƒƒƒƒƒƒƒƒƒƒƒƒƒƒƒƒƒƒƒƒƒƒƒƒƒƒƒƒƒƒƒƒƒƒƒƒ
```

```
          The SAS System          06:27 Sunday, November 4, 2012  2

Obs    _TYPE_    _FREQ_    xbar1    xbar2    p0val    betaprob    sd1

1         0         25     8.88     9.68     0.10       0.05      1.03

Obs  sd2   n1  n2    se            w1          w2        dfe      lowlim

1    1.00  25  25  0.27419  0.010955  .009891787   47.9688   0.62520
```

- JMP Data Table and formulas (Figures 2.8, 2.9, and 2.10)

Test 2.4 Single Mean (Two-Sided)

Parameters:
μ = population mean
σ = population standard deviation

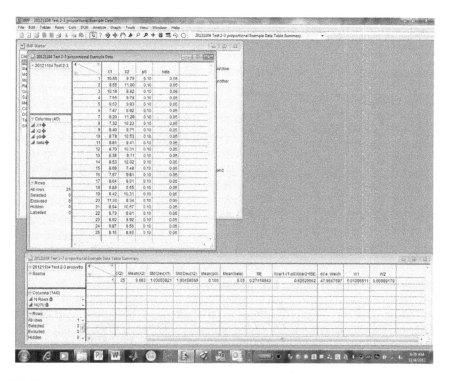

FIGURE 2.8
Test 2.3, JMP screen 1.

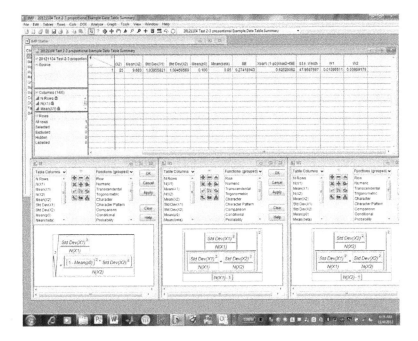

FIGURE 2.9
Test 2.3, JMP screen 2.

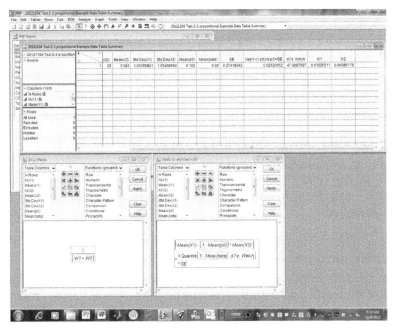

FIGURE 2.10
Test 2.3, JMP screen 3.

μ_L = lower acceptable limit for μ (L for "low")
μ_H = upper acceptable limit for μ (H for "high")
n = sample size
$1 - \beta$ = power to reject the null if $\mu = \mu_L$ or $\mu = \mu_H$

Hypotheses:

$$H_0: \mu < \mu_L \quad \text{OR} \quad \mu > \mu_H$$

$$H_1: \mu_L \leq \mu \leq \mu_H$$

Data:

$$\bar{X} = \text{sample mean}$$

$$S = \text{sample standard deviation}$$

$$n = \text{sample size}$$

Critical value(s):
  If

$$\bar{X} + t_{1-\beta}\frac{S}{\sqrt{n}} \geq \mu_L$$

and

$$\bar{X} - t_{1-\beta}\frac{S}{\sqrt{n}} \leq \mu_H$$

where $t_{1-\beta} = 100*(1 - \beta)$ percentile of a central t-distribution with $n - 1$ degrees of freedom, then reject H_0.

Discussion:
    This is an implementation of the two one-sided test, or TOST, philosophy of schuirmann (1987). That is, we do not split β in half, but apply all of this risk to each side of the hypothetical interval, (μ_L, μ_H). The reasoning for using TOST is twofold:

1. The "OR" in the null hypothesis statement is an exclusive "or." That is, μ cannot be both less than μ_L and greater than μ_H.
2. Inasmuch as failing to reject the null is not the desired state, reducing β by splitting it in half would be a less conservative criterion than not splitting the risk.

As a result, the power calculations are identical to those of the single mean (one-sided) test.

<u>Example:</u>
Suppose

$$\bar{X} = 10.0$$

$$S = 2.0$$

$$n = 25$$

$$\mu_L = 8.0$$

$$\mu_H = 9.5$$

$$\beta = 0.05$$

<u>Then:</u>

$$\bar{X} + t_{1-\beta}\frac{S}{\sqrt{n}} = 10.0 + 1.708 * \frac{2.0}{\sqrt{25}} \approx 10.68 > 8.0$$

$$\bar{X} - t_{1-\beta}\frac{S}{\sqrt{n}} = 10.0 - 1.708 * \frac{2.0}{\sqrt{25}} \approx 9.32 < 9.5$$

Therefore, we reject H_0 in favor of H_1: $8.0 \le \mu \le 9.5$, even though $\bar{X} = 10.0 > 9.5$.
Under TOST, the power calculations for the two-sided test for a single mean are like those for the one-sided test. However, instead of simply calculating

$$\Pr\left\{\bar{X} + t_{1-\beta}\frac{S}{\sqrt{n}} \ge \mu_L \mid \mu = \mu_a < \mu_L, n\right\} = \Pr\left\{\frac{\sqrt{n}(X-\mu_L)}{S} \ge -t_{1-\beta} \mid \mu = \mu_a < \mu_L, n\right\}$$

$$= 1 - \beta$$

we must also calculate

$$\Pr\left\{\bar{X} - t_{1-\beta}\frac{S}{\sqrt{n}} \le \mu_H \mid \mu = \mu_a > \mu_H, n\right\} = \Pr\left\{\frac{\sqrt{n}(X-\mu_H)}{S} \le +t_{1-\beta} \mid \mu = \mu_a > \mu_H, n\right\}$$

$$= 1 - \beta.$$

As in the one-sided case, the quantity:

$$\frac{\bar{X} - \mu_L}{S/\sqrt{n}}$$

has a noncentral t-distribution with noncentrality parameter

$$\delta_L = \frac{\sqrt{n}\,(\mu_a - \mu_L)}{\sigma}.$$

Similarly, the random variable

$$\frac{\overline{X} - \mu_H}{S/\sqrt{n}}$$

has a noncentral *t*-distribution with noncentrality parameter

$$\delta_H = \frac{\sqrt{n}\,(\mu_a - \mu_H)}{\sigma}.$$

Note that as either

$$\frac{\delta_L}{\sqrt{n}} = \frac{\mu_a - \mu_L}{\sigma} \quad \text{or} \quad \frac{\delta_H}{\sqrt{n}} = \frac{\mu_a - \mu_H}{\sigma}$$

goes to zero, the likelihood of rejecting the null hypothesis goes increasingly to $1 - \beta$.

Confidence interval formulation:
    The interval

$$\left(\overline{X} - t_{1-\beta}\frac{S}{\sqrt{n}}, \overline{X} + t_{1-\beta}\frac{S}{\sqrt{n}} \right)$$

is a $100(1 - 2\beta)$ percent confidence interval for μ. Using the example data yields the 90 percent confidence interval (9.32, 10.68).

Computational considerations:
    The code for this test is essentially the same as the code for Test 2.1, except that both a lower and upper test must be performed. In light of the TOST philosophy, both lower and upper test statistics should use a $100(1 - \beta)$ percentile from the appropriate *t*-distribution.

Test 2.5 A Special Case of Single Mean (Two-Sided)—Regression Slope

A special case of the single mean, two-sided test (Test 2.4) is for simple linear regression slope parameters. A common hypothesis and its alternate are:

$$H_0\!: \beta_1 < 1 - \Delta \quad \text{OR} \quad \beta_1 > 1 + \Delta$$

$$H_1: 1 - \Delta \le \beta_1 \le 1 + \Delta$$

where β_1 is the slope parameter and Δ is a number between 0 and 1.
    The standard error of the ordinary least squares (OLS) slope estimate is

$$SE = \frac{\hat{\sigma}}{\sqrt{\sum_{i=1}^{n}(X_i - \bar{X})^2}}$$

where $\hat{\sigma}$ is the root mean square error from the ANOVA for the regression.
The null hypothesis, H_0, is rejected if

$$b_1 + t_{1-\beta}SE \ge 1 - \Delta \quad \text{and} \quad b_1 - t_{1-\beta}SE \le 1 + \Delta$$

where b_1 is the ordinary least squares (OLS) slope estimate, and $t_{1-\beta} = 100*(1 - \beta)$ percentile of a central t-distribution with $n - 2$ degrees of freedom.

$$(b_1 - t_{1-\beta}SE, \; b_1 + t_{1-\beta}SE)$$

is a $100(1 - 2\beta)$ percent confidence interval for the slope parameter β_1 (do not confuse the slope β_1 with the Type II error risk, β).

Computational considerations:
    The computations for Test 2.5 are very similar to those for Test 2.4. The principal differences are a special case of the hypotheses, and the computation of the standard error of the estimate.

- SAS code

```
libname stuff 'H:\Personal Data\Equivalence & Noninferiority\
Programs & Output';

data calc;
  set stuff.d20121105_test_2_5_example_data;
  run;

proc means data = calc;

  var X Y del0 beta;
  output out = outmeans MEAN = xbar ybar delbar betabar N = nx
ny CSS = ssx ssy;
  run;

proc print data = calc;
  run;

proc print data = outmeans;
  run;
```

```
proc reg data = calc outest = outreg1;

model Y = X;

run;

data outreg2;
  set outreg1;
  drop _TYPE_;
  run;

data bigcalc;

set outmeans;
set outreg2;
seslope = _RMSE_/sqrt(ssx);
tval = tinv(1-betabar,nx-2);
lowlim = X + tval*seslope;
upplim = X - tval*seslope;
run;

proc print data = bigcalc;/*dataset bigcalc: var X is slope
estimate */
  var nx X lowlim upplim;
  run;
```

The output parameter "CSS" in proc reg is the corrected sums of squares. So, "ssx" is the corrected sums of squares for the X variable, as X is the first column in the data file.

The SAS System   09:33 Monday, November 5, 2012 10

The MEANS Procedure

Variable	Label	N	Mean	Std Dev	Minimum	Maximum
X	X	30	10.2108583	1.2373869	7.6748992	13.4843440
Y	Y	30	15.4861972	1.3359807	12.6635617	18.7004035
del0	del0	30	0.0750000	0	0.0750000	0.0750000
beta	beta	30	0.0500000	0	0.0500000	0.0500000

The SAS System   09:33 Monday, November 5, 2012 11

Obs	X	Y	del0	beta
1	10.05	15.12	0.075	0.05
2	8.97	14.55	0.075	0.05
3	9.78	15.62	0.075	0.05
4	7.67	12.66	0.075	0.05

```
 5   11.41   16.44   0.075   0.05
 6   10.69   16.05   0.075   0.05
 7   10.58   16.35   0.075   0.05
 8    9.39   14.16   0.075   0.05
 9    9.95   15.30   0.075   0.05
10   13.48   18.70   0.075   0.05
11    9.42   15.02   0.075   0.05
12    9.21   14.59   0.075   0.05
13   10.81   16.02   0.075   0.05
14   11.53   16.76   0.075   0.05
15   10.98   16.60   0.075   0.05
16   11.79   16.90   0.075   0.05
17   10.79   16.31   0.075   0.05
18   11.89   17.48   0.075   0.05
19    9.43   14.65   0.075   0.05
20   10.37   15.32   0.075   0.05
21    9.42   14.57   0.075   0.05
22   11.03   16.28   0.075   0.05
23    8.41   13.70   0.075   0.05
24   10.28   15.29   0.075   0.05
25   11.27   16.72   0.075   0.05
26    9.24   13.73   0.075   0.05
27    9.20   13.97   0.075   0.05
28   10.95   16.59   0.075   0.05
29   10.15   15.69   0.075   0.05
30    8.17   13.42   0.075   0.05
```

The SAS System   09:33 Monday, November 5, 2012 12

```
Obs _TYPE_ _FREQ_ xbar ybar delbar betabar nx ny   ssx      ssy

 1    0     30   10.21 15.49 0.075   0.05   30 30 44.4027 51.7605
```

The SAS System   09:33 Monday, November 5, 2012 13

The REG Procedure
Model: MODEL1
Dependent Variable: Y Y

Number of Observations Read   30
Number of Observations Used   30

Analysis of Variance

Source	DF	Sum of Squares	Mean Square	F Value	Pr > F
Model	1	49.05940	49.05940	508.56	<.0001
Error	28	2.70109	0.09647		
Corrected Total	29	51.76049			

```
Root MSE              0.31059   R-Square   0.9478
Dependent Mean       15.48620   Adj R-Sq   0.9460
Coeff Var             2.00561

                    Parameter Estimates

                       Parameter  Standard
Variable  Label    DF  Estimate      Error  t Value  Pr > |t|

Intercept Intercept  1   4.75325    0.47930     9.92    <.0001
X         X          1   1.05113    0.04661    22.55    <.0001

            The SAS System   09:33 Monday, November 5, 2012 14

         Obs    nx      X       lowlim    upplim

          1     30   1.05113   1.13042   0.97184
```

To reject H_0, lowlim must be greater than or equal to $1 - \Delta = 0.975$ and upplim must be less than or equal to $1 + \Delta = 1.075$.

- JMP Data Table and formulas (see Figures 2.11, 2.12, and 2.13)

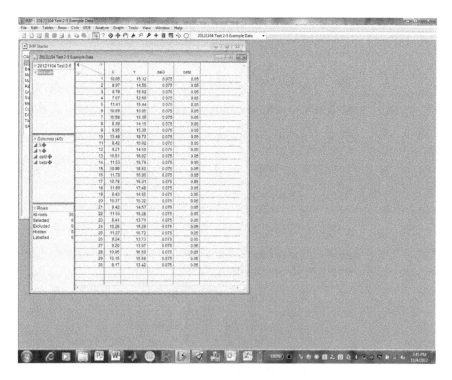

FIGURE 2.11
Test 2.5, JMP screen 1.

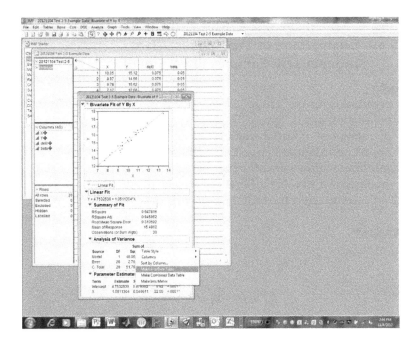

FIGURE 2.12
Test 2.5, JMP screen 2.

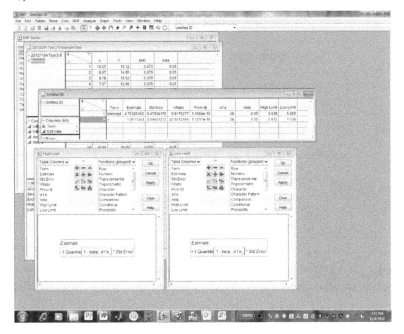

FIGURE 2.13
Test 2.5, JMP screen 3.

Test 2.6.1 Comparison of Two Means—Two Independent Samples (Two-Sided)

Parameters:

μ_1 = population mean, "group" 1
μ_2 = population mean, "group" 2
σ_1 = population standard deviation, "group" 1
σ_2 = population standard deviation, "group" 2
Δ_L = minimum allowable difference between μ_1 and μ_2
Δ_H = maximum allowable difference between μ_1 and μ_2
$1 - \beta$ = power to reject the null if $\mu_1 - \mu_2 = \Delta_L$ or $\mu_1 - \mu_2 = \Delta_H$

Hypotheses:

$$H_0: \mu_1 - \mu_2 < \Delta_L \quad \text{OR} \quad \mu_1 - \mu_2 > \Delta_H$$

$$H_1: \Delta_L \leq \mu_1 - \mu_2 \leq \Delta_H$$

Data:

$$\bar{X}_1 = \text{sample mean, "group" 1}$$

$$S_1 = \text{sample standard deviation, "group" 1}$$

$$n_1 = \text{sample size, "group" 1}$$

$$\bar{X}_2 = \text{sample mean, "group" 2}$$

$$S_2 = \text{sample standard deviation, "group" 2}$$

$$n_2 = \text{sample size, "group" 2}$$

Critical value(s):

Reject H_0 if:

$$\bar{X}_1 - \bar{X}_2 + t_{1-\beta}SE \geq \Delta_L \quad \text{and} \quad \bar{X}_1 - \bar{X}_2 - t_{1-\beta}SE \leq \Delta_H$$

where $t_{1-\beta}$ = 100*(1 − β) percentile of a central t-distribution with $n_1 + n_2 - 2$ degrees of freedom and SE is the standard error for the difference of two means:

$$SE = \sqrt{\frac{S_1^2}{n_1} + \frac{S_2^2}{n_2}}.$$

Discussion:

There are several special cases of the comparison of two means. Comparison of regression slopes is such a case. Assume that slopes forming two independent simple linear regressions are to be compared, where the response variable was measured under two competing methods and the regressor is the same for both models. The two means are replaced with two slope estimates. The standard error for each slope is

$$SE(b_1) = \frac{s}{\sqrt{\sum_{i=1}^{n}(X_i - \bar{X})^2}}$$

where s is the root mean square error from the regression:

$$Y_i = \beta_0 + \beta_1 X_i + \varepsilon_i$$

$i = 1, n$, and b_1 is the slope estimate. If there are two slopes to be compared, say, β_1 and γ_1, and the hypotheses are

$$H_0: \beta_1 - \gamma_2 < \Delta_L \quad \text{OR} \quad \beta_1 - \gamma_2 > \Delta_H$$

$$H_1: \Delta_L \leq \beta_1 - \gamma_2 \leq \Delta_H$$

with slope estimates b_1 and c_1, the rejection rule is

Reject H_0 if:

$$b_1 - c_2 + t_{1-\beta} SE \geq \Delta_L, \quad \text{and} \quad b_1 - c_2 - t_{1-\beta} SE \leq \Delta_H$$

where $t_{1-\beta} = 100*(1 - \beta)$ percentile of a central t-distribution with $dfe_1 + dfe_2$ degrees of freedom, where dfe_1 and dfe_2 are the respective degrees of freedom of error for the two independent regressions, and SE is given by:

$$SE = \sqrt{SE(b_1)^2 + SE(c_1)^2}.$$

As in the case of all two-sided equivalence tests, we have adopted the TOST philosophy for the two-sided comparison of two means or slopes.

Example:

Suppose we hypothesize:

$$H_0: \mu_1 - \mu_2 < -5.0 \quad \text{OR} \quad \mu_1 - \mu_2 > +5.0$$

$$H_1: -5.0 \leq \mu_1 - \mu_2 \leq +5.0$$

The data are

$$\bar{X}_1 = 96.0$$

$$S_1 = 2.40$$

$$n_1 = 12$$

$$\bar{X}_2 = 101.5$$

$$S_2 = 2.20$$

$$n_2 = 14$$

$$\bar{X}_1 - \bar{X}_2 = 96 - 101.5 = -5.5.$$

We choose $1 - \beta = 0.95$, so $t_{1-\beta} = 1.711$ ($12 + 14 - 2 = 24$ degrees of freedom). We compute

$$SE = \sqrt{\frac{S_1^2}{n_1} + \frac{S_2^2}{n_2}} = \sqrt{\frac{2.40^2}{12} + \frac{2.20^2}{14}} \approx 0.909.$$

The critical values are

$$\bar{X}_1 - \bar{X}_2 + t_{1-\beta}SE = 96 - 101.5 + 1.711 * 0.909 \approx -3.94 \geq -5.0$$

and

$$\bar{X}_1 - \bar{X}_2 - t_{1-\beta}SE = 96 - 101.5 - 1.711 * 0.909 \approx -7.05 \leq +5.0.$$

Therefore, we reject H_0.

The probability of rejecting the null hypothesis when $\mu_1 - \mu_2 = \Delta_L$ is

$$\Pr\left\{\bar{X}_1 - \bar{X}_2 + t_{1-\beta}SE \geq \Delta_L\right\} = 1 - \beta$$

or in other words

$$\Pr\left\{\frac{\bar{X}_1 - \bar{X}_2 - \Delta_L}{SE} \geq -t_{1-\beta}\right\} = 1 - \beta$$

Similarly, the probability of rejecting the null hypothesis when $\mu_1 - \mu_2 = \Delta_H$ is

$$\Pr\left\{\bar{X}_1 - \bar{X}_2 - t_{1-\beta}SE \leq \Delta_H\right\} = 1 - \beta.$$

Under some specific alternate hypothesis, such as

$$\mu_1 - \mu_2 = \Delta_a$$

where $\Delta_a < \Delta_L$,

$$t = \frac{\bar{X}_1 - \bar{X}_2 - \Delta_L}{\sqrt{\dfrac{S_1^2}{n_1} + \dfrac{S_2^2}{n_2}}}$$

has a noncentral t-distribution with $n_1 + n_2 - 2$ degrees of freedom and non-centrality parameter

$$\delta = \frac{\Delta_a - \Delta_L}{\sqrt{\sigma_1^2/n_1 + \sigma_2^2/n_2}}.$$

As $\delta \to 0$, the probability of rejecting H_0 goes to $1 - \beta$. A similar statement can be made for $\Delta_a > \Delta_H$.

Confidence interval formulation:

$$\left(\bar{X}_1 - \bar{X}_2 - t_{1-\beta}SE, \bar{X}_1 - \bar{X}_2 + t_{1-\beta}SE\right)$$

is a $100^*(1 - 2\beta)$ percent confidence interval for $\mu_1 - \mu_2$. In the example, the 90 percent confidence interval for $\mu_1 - \mu_2$ is $(-7.05, -3.94)$.

Computational considerations:
    The code for this test is essentially the same as the code for Test 2.2, except that both a lower and upper test must be performed. In light of the TOST philosophy, both lower and upper test statistics should use a $100(1 - \beta)$ percentile from the appropriate t-distribution.

Test 2.6.2 A Special Case of Comparison of Two Means—Two Independent Samples (Two-Sided), Two-Level Factorial Validation/Verification Experiment

Consider a situation in which a process has multiple "factors" that affect the distribution of some quality characteristic of the product of the process. Assume that the quality characteristic is a continuously valued random variable, and that the factor levels can be controlled, at least in an experimental setting. Furthermore, suppose that for each factor, two levels can be identified as being lower and upper bounds, respectively, on the operating level for the factor. In this discussion, we treat each factor as having the characteristic that it can be set to any value over a continuum, ranging from a lowest level (L) to a highest level (H). While the assumption of a continuously valued factor is not absolutely necessary for the statistical properties to hold true, it will simplify the illustrations of those properties.

The objective of such an experiment is to show that within the low and high ranges of all the factors (called the operating space) the average value of the quality characteristic does not change "enough to matter."

It is assumed that each factor will vary between exactly two levels for the experiment. Each combination of conditions under which the experiment is performed is called a "run." It is assumed that for each run, a fixed number of observations, or measurements, of the quality characteristic, will be made. We will refer to each observation as a "replicate," although it is possible that the observations are not "true" replicates, in that they are all observed within a single "run."

Parameters:

k = the number of factors.

2^{-p} = the fraction of the total number of possible combinations of factor levels to be used in the experiment (p could be 0, so that all possible combinations will be included, i.e., a full factorial experiment).

n = number of replicates for each run.

$\mu_i^{(L)}$ = the average value of the quality characteristic (also called the response variable) when the ith factor is set to its low level (L).

$\mu_i^{(H)}$ = the average value of the quality characteristic (also called the response variable) when the ith factor is set to its high level (H).

$E_i = \mu_i^{(H)} - \mu_i^{(L)}$, the effect of the ith factor on the average value of the quality characteristic.

$\hat{\sigma}$ = the estimate of the within-run standard deviation of the quality characteristic, estimated using the square root of the mean square error term from an ANOVA. It is assumed that the within-run "population" standard deviation, σ, is the same regardless of the run conditions.

Δ_L = the lower bound on the tolerable effect of the ith factor.

Δ_H = the upper bound on the tolerable effect of the ith factor.

Hypotheses:

$$H_0: E_i < \Delta_L \quad \text{OR} \quad E_i > \Delta_H$$

$$H_1: \Delta_L \le E_i \le \Delta_H$$

Data:

$$\hat{E}_i = \hat{\mu}_i^H - \mu_i^L = \text{the estimated effect of the } i\text{th factor}$$

$$SE = \frac{2\hat{\sigma}}{\sqrt{n2^{k-p}}} = \text{the standard error of the effect estimate}$$

Critical value(s):
Reject H_0 if:

$$\hat{E}_i + t_{1-\beta}SE \ge \Delta_L \quad \text{and} \quad \hat{E}_i - t_{1-\beta}SE \le \Delta_H$$

where $t_{1-\beta} = 100*(1 - \beta)$ percentile of a central t-distribution with $n2^{k-p} - T$ degrees of freedom, where T is the number of terms in the ANOVA and SE is the standard error of the estimated effect.

Discussion:
The power calculations are very similar to those for the comparison of two means, two-sided test. The quantity:

$$\frac{\hat{E}_i - \Delta_L}{SE}$$

has a noncentral t-distribution with noncentrality parameter

$$\delta = \frac{\Delta_a - \Delta_L}{\dfrac{2\sigma}{\sqrt{n2^{k-p}}}} = \frac{(\Delta_a - \Delta_L)\sqrt{n2^{k-p}}}{2\sigma}$$

or

$$\frac{\delta}{\sqrt{n}} = \frac{(\Delta_a - \Delta_L)\sqrt{2^{k-p}}}{2\sigma}.$$

Similar to the case of comparing two means, as $\delta \to 0$, the probability of rejecting H_0 goes to $1 - \beta$.

Example:
Consider a full two-factor design ($k = 2, p = 0$), with $n = 2$ replicates per run (Table 2.1), with $\Delta_L = -1.0$ and $\Delta_H = +1.0$, and $1 - \beta = 0.95$.

TABLE 2.1

2^4 Full Factorial Experiment with $n = 2$ Replicates

Run	Factor A	Factor B	Replicate	Response
1	Low	Low	1	10.1
1	Low	Low	2	10.5
2	Low	High	1	12.3
2	Low	High	2	12.7
3	High	Low	1	8.1
3	High	Low	2	8.3
4	High	High	1	16.9
4	High	High	2	16.8

From these data:

$$\hat{E}_A = \hat{\mu}_A^H - \mu_A^L \approx 1.125$$

$$SE \approx 0.15207$$

$$t_{1-\beta} \approx 2.1318$$

$$\hat{E}_A + t_{1-\beta}SE \approx 1.45 \geq \Delta_L$$

$$\hat{E}_A - t_{1-\beta}SE \approx 0.80 \leq \Delta_H$$

Therefore, we reject H_0 and conclude that factor A does not have a meaningful effect on the response.

Confidence interval formulation:

$$\left(\hat{E}_A - t_{1-\beta}SE, \hat{E}_A + t_{1-\beta}SE \right)$$

is a $100*(1 - 2\beta)$ percent confidence interval for the effect of factor A. The example data yield a 90 percent confidence interval of (0.80, 1.45) for the effect of factor A.

Computational considerations:
    In the example used to illustrate the computer code and output, a full 2^3 factorial experiment with $n = 1$ replicate was used. The three-way interaction term was not included, which provides a single-degree-of-freedom error term. In general, it is not recommended to use an experiment with such a low power for validation. The example presented here is only used to illustrate the computing methods.

- SAS code

```
libname stuff 'H:\Personal Data\Equivalence & Noninferiority\
Programs & Output';

data calc;
   set stuff.d20121107_test_2_6_2_example_dat;

   run;

/* this proc glm will produce two-one-sided 95% confidence
limits for the effect of X1 */
proc glm data = calc;
   class X1 X2 X3;
   model Y = X1 X2 X3 X1*X2 X1*X3 X2*X3/ALPHA = 0.10 CLPARM;/*
   note alpha set to 0.10 */
   estimate 'X1 effect' X1 -1 1;
```

```
  lsmeans X1/out = outcalc;/* this output file provides an
alternate way of making the calculations */
  run;

data comp1;
  set outcalc;
  prod = X1 * LSMEAN;
  SESQ = STDERR**2;
  run;

proc means data = comp1;
  var prod sesq;
  output out = comp2 SUM = effect sumsq;
  run;

data outcomp;
  set comp2;
  beta = 0.05;
  sediff = sqrt(sumsq);
  lowlim = effect - tinv(1-beta,1)*sediff;
  upplim = effect + tinv(1-beta,1)*sediff;
  run;

proc print data = calc;/*dataset calc has input data*/
  run;

proc print data = outcomp;/*effect with std error of effect
and limits */
  run;
```

The GLM Procedure

Class Level Information

Class	Levels	Values
X1	2	-1 1
X2	2	-1 1
X3	2	-1 1

Number of Observations Read   8
Number of Observations Used   8
The SAS System 05:14 Wednesday, November 7, 2012 43

The GLM Procedure

Dependent Variable: Y Y

Source	DF	Sum of Squares	Mean Square	F Value	Pr > F
Model	6	43.76912169	7.29485362	287.62	0.0451

```
Error               1    0.02536297    0.02536297

Corrected Total     7   43.79448466

              R-Square    Coeff Var    Root MSE      Y Mean

              0.999421    -87.62644    0.159258    -0.181746

Source     DF      Type I SS    Mean Square    F Value    Pr > F

X1          1    5.22305892    5.22305892     205.93    0.0443
X2          1   37.59467830   37.59467830    1482.27    0.0165
X3          1    0.00971073    0.00971073       0.38    0.6472
X1*X2       1    0.49558028    0.49558028      19.54    0.1416
X1*X3       1    0.37802734    0.37802734      14.90    0.1614
X2*X3       1    0.06806613    0.06806613       2.68    0.3489

Source     DF    Type III SS    Mean Square    F Value    Pr > F

X1          1    5.22305892    5.22305892     205.93    0.0443
X2          1   37.59467830   37.59467830    1482.27    0.0165
X3          1    0.00971073    0.00971073       0.38    0.6472
X1*X2       1    0.49558028    0.49558028      19.54    0.1416
X1*X3       1    0.37802734    0.37802734      14.90    0.1614
X2*X3       1    0.06806613    0.06806613       2.68    0.3489

                        Standard
Parameter  Estimate      Error    t Value Pr > |t| 90% Confidence Limits
X1 effect 1.61602273   0.11261210  14.35    0.0443  0.90501791  2.32702755

          The SAS System 05:14 Wednesday, November 7, 2012 44

                    The GLM Procedure
                    Least Squares Means

              X1          Y LSMEAN

              -1          -0.98975737
               1           0.62626536

          The SAS System 05:14 Wednesday, November 7, 2012 45

                    The MEANS Procedure

    Variable  N      Mean    Std Dev    Minimum    Maximum
    ƒƒƒƒƒƒƒƒƒƒƒƒƒƒƒƒƒƒƒƒƒƒƒƒƒƒƒƒƒƒƒƒƒƒƒƒƒƒƒƒƒƒƒƒƒƒƒƒƒƒ
    prod      2  0.8080114  0.2570277  0.6262654  0.9897574
    SESQ      2  0.0063407          0  0.0063407  0.0063407
    ƒƒƒƒƒƒƒƒƒƒƒƒƒƒƒƒƒƒƒƒƒƒƒƒƒƒƒƒƒƒƒƒƒƒƒƒƒƒƒƒƒƒƒƒƒƒƒƒƒƒ
```

```
The SAS System 05:14 Wednesday, November 7, 2012 46

        Obs   X1    X2    X3      Y    delL   delH

         1     1     1     1    2.58   -1.5   1.5
         2     1     1    -1    2.51   -1.5   1.5
         3     1    -1     1   -0.96   -1.5   1.5
         4     1    -1    -1   -1.62   -1.5   1.5
         5    -1     1     1    1.14   -1.5   1.5
         6    -1     1    -1    1.72   -1.5   1.5
         7    -1    -1     1   -3.62   -1.5   1.5
         8    -1    -1    -1   -3.19   -1.5   1.5


    The SAS System 05:14 Wednesday, November 7, 2012 47

Obs _TYPE_  _FREQ_  effect    sumsq   beta sediff  lowlim  upplim

 1    0        2    1.61602 0.012681 0.05 0.11261 0.90502 2.32703
```

- JMP Data Table and formula

  To obtain the same results with JMP as SAS, the factors X1, X2, and X3 must be modeling type "nominal" (Figures 2.14, 2.15, 2.16. and 2.17).

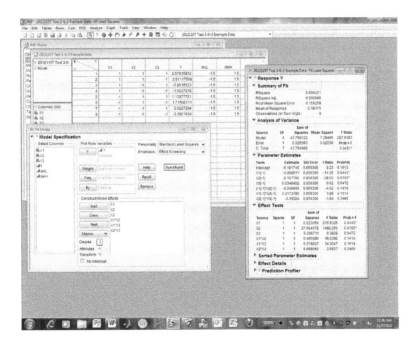

FIGURE 2.14
Test 2.6.2, JMP screen 1.

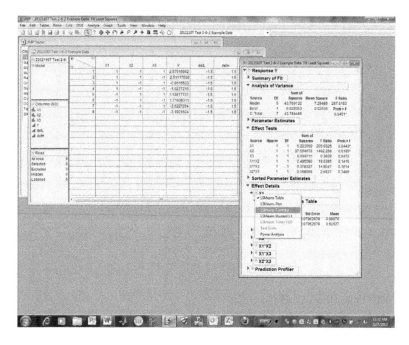

FIGURE 2.15
Test 2.6.2, JMP screen 2.

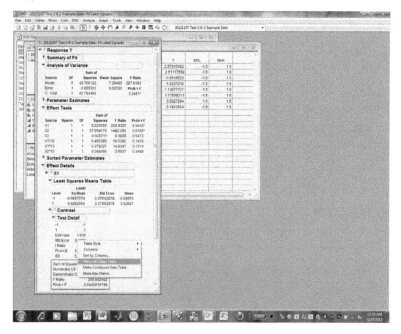

FIGURE 2.16
Test 2.6.2, JMP screen 3.

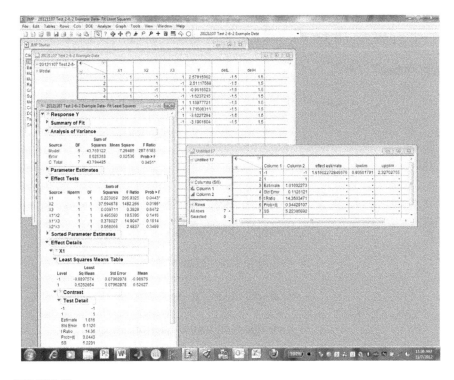

FIGURE 2.17
Test 2.6.2, JMP screen 4.

Since for factor X1, estimate $-tSE \approx 0.905 \leq \Delta_U = +1.5$ and estimate $+tSE \approx 2.327 \geq \Delta_L = -1.5$, conclude that the low (–1) and high (+1) levels of X1 are equivalent with respect to response variable Y.

The value of t in this case in the $100(1 - \beta)$ percent = the 95th percentile of a t-distribution with 1 degree of freedom (based on the error term for the linear model).

Test 2.7 Comparison of Two Means—Proportional Difference Paradigm for Two Independent Samples (Two-Sided)

Parameters:
  μ_1 = population mean, "group" 1
  μ_2 = population mean, "group" 2
  σ_1 = population standard deviation, "group" 1
  σ_2 = population standard deviation, "group" 2
  p_L = minimum allowable proportion difference between μ_1 and μ_2.

p_H = maximum allowable proportion difference between μ_1 and μ_2.
$1 - \beta$ = power to reject the null if $\mu_1 - (1 - p_L)\mu_2 = 0$ or $\mu_1 - (1 + p_H)\mu_2 = 0$

<u>Hypotheses</u>:

$$H_0: \mu_1 - (1 - p_L)\mu_2 < 0 \quad \text{OR} \quad \mu_1 - (1 + p_H)\mu_2 > 0$$

$$H_1: (1 - p_L)\mu_2 \leq \mu_1 \leq (1 + p_H)\mu_2$$

<u>Data</u>:

$$\bar{X}_1 = \text{sample mean, "group" 1}$$

$$S_1 = \text{sample standard deviation, "group" 1}$$

$$n_1 = \text{sample size, "group" 1}$$

$$\bar{X}_2 = \text{sample mean, "group" 2}$$

$$S_2 = \text{sample standard deviation, "group" 2}$$

$$n_2 = \text{sample size, "group" 2}$$

<u>Critical value(s)</u>:
  Reject H_0 if:

$$\bar{X}_1 - (1 - p_L)\bar{X}_2 + t_{1-\beta}SE_L \geq 0$$

and

$$\bar{X}_1 - (1 + p_H)\bar{X}_2 - t_{1-\beta}SE_H \leq 0$$

where $t_{1-\beta} = 100*(1 - \beta)$ percentile of a central t-distribution with $n_1 + n_2 - 2$ degrees of freedom, and SE_L is the standard error for the difference $\bar{X}_1 - (1 - p_L)\bar{X}_2$:

$$SE_L = \sqrt{\frac{S_1^2}{n_1} + \frac{(1 - p_L)^2 S_2^2}{n_2}}$$

and SE_H is the standard error for the difference $\bar{X}_1 - (1 + p_H)\bar{X}_2$:

$$SE_H = \sqrt{\frac{S_1^2}{n_1} + \frac{(1 + p_H)^2 S_2^2}{n_2}}.$$

Discussion:

While this TOST is very similar to the comparison of two means with maximum allowable differences Δ_L and Δ_H, the biggest difference is that there are two standard error formulas, depending on the side of the test (i.e., $1 - p_L$ or $1 + p_H$). Inasmuch as this is a TOST, the power calculations are done separately for each side of the test, and are therefore identical to those of the single-sided "proportional" difference paradigm.

Example:

Suppose we hypothesize:

$$H_0: \mu_1 - (1 - 0.01)\mu_2 < 0 \quad \text{OR} \quad \mu_1 - (1 + 0.01)\mu_2 > 0$$

$$H_1: (1 - 0.01)\mu_2 \leq \mu_1 \leq (1 + 0.01)\mu_2$$

In other words, we hope that μ_1 is in the interval $(0.99\mu_2, 1.01\mu_2)$. Note that in this example, $p_L = p_H = 0.01$.

The data are:

$$\bar{X}_1 = 98.9$$

$$S_1 = 2.40$$

$$n_1 = 12$$

$$\bar{X}_2 = 100.0$$

$$S_2 = 2.20$$

$$n_2 = 14$$

We choose $1 - \beta = 0.95$, so $t_{1-\beta} = 1.711$ ($12 + 14 - 2 = 24$ degrees of freedom). We compute

$$SE_L = \sqrt{\frac{S_1^2}{n_1} + \frac{(1 - p_L)^2 S_2^2}{n_2}} = \sqrt{\frac{2.40^2}{12} + \frac{(1 - 0.01)^2 2.2^2}{14}} \approx 0.9049$$

$$SE_H = \sqrt{\frac{S_1^2}{n_1} + \frac{(1 + p_H)^2 S_2^2}{n_2}} = \sqrt{\frac{2.40^2}{12} + \frac{(1 + 0.01)^2 2.2^2}{14}} \approx 0.9125.$$

Comparing the test statistics to the critical value of 0:

$$\bar{X}_1 - (1 - p_L)\bar{X}_2 + t_{1-\beta}SE_L = 98.9 - (1 - 0.01)100.0 + 1.711(0.9049) \approx 1.448 \geq 0.$$

and

$$\bar{X}_1 - (1 + p_H)\bar{X}_2 - t_{1-\beta}SE_H = 98.9 - (1 + 0.01)100.0 - 1.711(0.9125) \approx -3.661 \leq 0.$$

Therefore, we reject H_0.

Confidence interval formulation:

$$\bar{X}_1 - (1 - p_L)\bar{X}_2 + t_{1-\beta}SE_L$$

is an upper $100(1 - \beta)$ percent confidence limit on $\mu_1 - (1 - p_L)\mu_2$.

$$\bar{X}_1 - (1 + p_H)\bar{X}_2 - t_{1-\beta}SE_H$$

is a lower $100(1 - \beta)$ percent confidence limit on $\mu_1 - (1 + p_H)\mu_2$.

Thus, in the example, the 95 percent lower confidence limit for $\mu_1 - (1.01)\mu_2$ is -3.661. The 95 percent upper confidence limit for $\mu_1 - (0.99)\mu_2$ is 1.448.

Computational considerations:

The computer code for this test is very similar to that for Test 2.3. Both a lower and upper limit would be computed. Also, recall that the formulas for the lower and upper limit standard errors are different, in that the SE_L involves the factor $1 - p_L$ and the SE_H involves the factor $1 + p_H$.

3

Variances (Standard Deviations) and Coefficients of Variation

Test 3.1 Single Variance (One-Sided)

<u>Parameters:</u>
σ_0^2 = maximum tolerable variance
$1 - \beta$ = power to reject the null if variance equals σ_0^2

<u>Hypotheses:</u>

$$H_0: \sigma^2 > \sigma_0^2$$

$$H_1: \sigma^2 \le \sigma_0^2$$

<u>Data:</u>

$$S = \text{sample standard deviation} = \sqrt{\frac{1}{n-1}\sum_{i=1}^{n}\left(X_i - \bar{X}\right)^2}$$

<u>Critical value(s):</u>
  Reject H_0 if

$$S^2 \le \frac{\sigma_0^2}{n-1}\chi_{1-\beta}^2(n-1)$$

where $\chi_{1-\beta}^2(n-1) = 100(1-\beta)$ percentile of a chi-squared distribution with $n-1$ degrees of freedom.

<u>Discussion:</u>
  The quantity

$$\frac{(n-1)S^2}{\sigma_a^2}$$

has a chi-squared distribution when $\sigma^2 = \sigma_a^2$. Suppose that

$$\sigma_a^2 = k^2 \sigma_0^2 \quad \text{for } k > 1.$$

Then

$$\frac{(n-1)S^2}{\sigma_a^2} = \frac{(n-1)S^2}{k^2 \sigma_0^2}$$

and

$$\Pr\left\{\frac{(n-1)S^2}{\sigma_a^2} \le \chi_{1-\beta}^2(n-1) \,|\, \sigma^2 = \sigma_a^2\right\} = \Pr\left\{\frac{(n-1)S^2}{\sigma_0^2} \le \frac{\chi_{1-\beta}^2(n-1)}{k^2}\right\}.$$

For $k = 1$, this probability is $1 - \beta$. As k gets large, the probability, which is the power, decreases.

Example:
  Suppose: $\sigma_0^2 = 4.0$, $n = 20$, and $S = 2.1$. If $1 - \beta = 0.95$, then

$$\chi_{1-\beta}^2(n-1) \approx 30.144.$$

Since

$$S^2 = 4.41 \le \frac{\sigma_0^2}{n-1}\chi_{1-\beta}^2(n-1) = \frac{4}{19}\,30.144 \approx 6.35$$

we reject H_0. The power curve for this test is given in Figure 3.1 (power as a function of the multiplier, k).

Confidence interval formulation:

$$\hat{\sigma}_U = \sqrt{\frac{(n-1)S^2}{\chi_\beta^2(n-1)}}$$

is an upper $100(1 - \beta)$ percent confidence limit on σ. In the example,

$$\hat{\sigma}_U = \sqrt{\frac{(n-1)S^2}{\chi_\beta^2(n-1)}} = \sqrt{\frac{(20-1)4.41}{10.117}} \approx 2.878 \left(\chi_{0.05}^2(20-1) \approx 10.117\right).$$

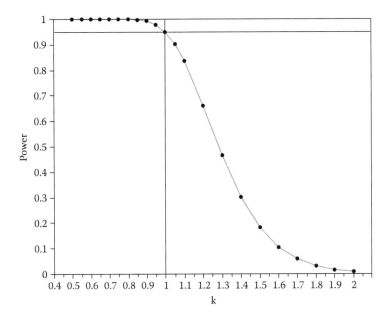

FIGURE 3.1
Test 3.1, power curve for testing a single variance.

Computational considerations:

- SAS code

```
libname stuff 'H:\Personal Data\Equivalence & Noninferiority\
Programs & Output';

data calc;

  set stuff.d20121107_test_3_1_example_data;

  run;

proc means data = calc;
  var X sig0 beta;
  output out = onemean MEAN = xbar popsig betaprob
  VAR = sampvar N = n1;
  run;

data outcalc;
  set onemean;
  upplim = (popsig**2)*cinv(1-betaprob,n1-1)/(n1-1);
  run;
```

```
proc print data = outcalc;/* has vars xbar popsig betaprob
sampvar n1 upplim */
  run;
```

The SAS System 13:53 Wednesday, November 7, 2012 2

                        The MEANS Procedure

Variable	Label	N	Mean	Std Dev	Minimum	Maximum
X	X	30	0.1613638	0.7395784	-1.5758867	1.5200489
sig0	sig0	30	1.2000000	0	1.2000000	1.2000000
beta	beta	30	0.0500000	0	0.0500000	0.0500000

        The SAS System 13:53 Wednesday, November 7, 2012 3

Obs	_TYPE_	_FREQ_	xbar	popsig	betaprob	sampvar	n1	upplim
1	0	30	0.161	1.200	0.05	0.54698	30	2.11317

- JMP Data Table and formulas (Figure 3.2)

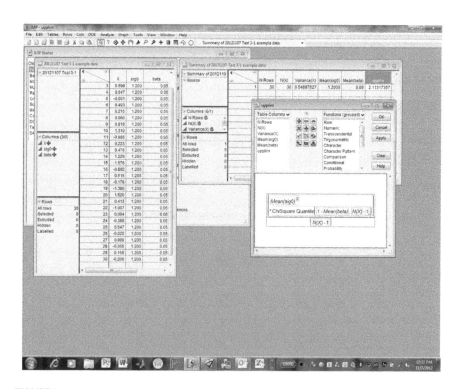

FIGURE 3.2
Test 3.1, JMP.

Test 3.2 Comparison of Two Variances (One-Sided)

Parameters:

σ_1^2 = variance of population 1

σ_2^2 = variance of population 2

k_0^2 = maximum tolerable ratio of the two variances

$$\frac{\sigma_1^2}{\sigma_2^2}$$

$1 - \beta$ = power to reject the null if

$$\frac{\sigma_1^2}{\sigma_2^2} = k_0^2.$$

Hypotheses:

$$H_0: \frac{\sigma_1^2}{\sigma_2^2} > k_0^2$$

$$H_1: \frac{\sigma_1^2}{\sigma_2^2} \le k_0^2$$

Data:

S_1 = sample standard deviation, sampled from population 1

S_2 = sample standard deviation, sampled from population 2

n_1 = sample size from population 1

n_2 = sample size from population 2

In general, the formula employed here for sample standard deviation is assumed to be

$$S = \sqrt{\frac{1}{n-1} \sum_{i=1}^{n} (X_i - \overline{X})^2}.$$

Critical value(s):

Reject H_0 if:

$$\frac{(n_1 - 1)S_1^2}{(n_2 - 1)S_2^2} \le \frac{1}{k_0^2} F_{1-\beta}(n_1 - 1, n_2 - 1).$$

Discussion:
  If

$$\sigma_1^2 = k_0^2 \sigma_2^2$$

then

$$\frac{(n_1 - 1)S_1^2 / k_0^2 \sigma_2^2}{(n_2 - 1)S_2^2 / \sigma_2^2}$$

has an F-distribution with $n_1 - 1$ and $n_2 - 1$ degrees of freedom in the numerator and denominator, respectively. Thus,

$$\frac{(n_1 - 1)S_1^2}{(n_2 - 1)S_2^2} \sim k_0^2 F(n_1 - 1, n_2 - 1).$$

Therefore,

$$\Pr\left\{ k_0^2 F(n_1 - 1, n_2 - 1) \le F_{1-\beta}(n_1 - 1, n_2 - 1) \right\}$$

$$= \Pr\left\{ F(n_1 - 1, n_2 - 1) \le \frac{1}{k_0^2} F_{1-\beta}(n_1 - 1, n_2 - 1) \right\}$$

$$= 1 - \beta$$

so

$$\Pr\left\{ \frac{(n_1 - 1)S_1^2}{(n_2 - 1)S_2^2} \le \frac{1}{k_0^2} F_{1-\beta}(n_1 - 1, n_2 - 1) \right\} = 1 - \beta.$$

  If in fact

$$\sigma_1^2 = k_a^2 \sigma_2^2, \quad k_a > k_0$$

then, as k_a^2 gets larger than k_0^2, the power decreases. That is, the power is given by:

$$\Pr\left\{ F \le \frac{k_0^2}{k_a^2} F_{1-\beta}(n_1 - 1, n_2 - 1) \right\}.$$

Example:
  Suppose:
  $k_0^2 = 1.0$

$S_1 = 5.0$
$S_2 = 6.0$
$n_1 = n_2 = 10$
and $1 - \beta = 0.95$, so

$$\frac{(n_1 - 1)S_1^2}{(n_2 - 1)S_2^2} = \frac{9 * 5.0^2}{9 * 6.0^2} \approx 0.694$$

$$\frac{1}{k_0^2} F_{1-\beta}(n_1 - 1, n_2 - 1) \approx 3.1789.$$

Since

$$\frac{(n_1 - 1)S_1^2}{(n_2 - 1)S_2^2} \leq \frac{1}{k_0^2} F_{1-\beta}(n_1 - 1, n_2 - 1) \approx 3.1789$$

we reject H_0.

The power curve for this example is given as a function of $k_a \geq k_0 = 1.0$ in Figure 3.3.

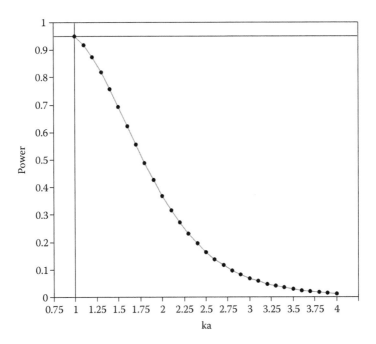

FIGURE 3.3
Test 3.2, power curve for comparison of two variances.

<u>Confidence interval formulation:</u>

$$\hat{k}_U = \sqrt{\frac{(n_2-1)S_2^2}{(n_1-1)S_1^2} F_{1-\beta}(n_1-1, n_2-1)}$$

is an upper $100(1-\beta)$ percent confidence limit for

$$k = \frac{\sigma_1}{\sigma_2}.$$

The example data yield:

$$\hat{k}_U = \sqrt{\frac{(n_2-1)S_2^2}{(n_1-1)S_1^2} F_{1-\beta}(n_1-1, n_2-1)} = \sqrt{\frac{(10-1)36}{(10-1)25}(3.1789)} \approx 4.578.$$

Note that as n_1 and n_2 get large, $F_{1-\beta}(n_1-1, n_2-1)$ will get smaller.

<u>Computational considerations:</u>

- SAS code

```
libname stuff 'H:\Personal Data\Equivalence & Noninferiority\
Programs & Output';

data calc;

   set stuff.d20121107_test_3_2_example_data;

   run;

proc means data = calc;
   var X1 X2 k0;
   output out = onemean MEAN = xbar1 xbar2 k0val VAR = v1 v2
   N = n1 n2;
   run;

data outcalc;
   set onemean;
   beta = 0.05;
   var_ratio = ((n1-1)*v1)/((n2-1)*v2);
   crit_val = finv(1-beta,n1-1,n2-1)/k0val**2;
   run;

proc print data = outcalc;/* has vars xbar1 xbar2 k0val v1 v2
n1 n2 beta var_ratio crit_val */
   run;
```

```
        The SAS System 13:53 Wednesday, November 7, 2012 7

                        The MEANS Procedure

Variable Label   N       Mean      Std Dev    Minimum      Maximum
ffffffffffffffffffffffffffffffffffffffffffffffffffffffffffffffffffff
X1         X1     35  0.1920832  1.8406642  -3.8176650   4.3223639
X2         X2     35  0.1100396  2.0376728  -4.5550806   4.7211140
k0         k0     35  1.0500000          0   1.0500000   1.0500000
ffffffffffffffffffffffffffffffffffffffffffffffffffffffffffffffffffff

        The SAS System 13:53 Wednesday, November 7, 2012 8

                                                          var_
Obs _TYPE_ _FREQ_ xbar1 xbar2 k0val   v1      v2    n1 n2 beta  ratio   crit_val
1      0      35   0.192 0.110  1.05 3.38804 4.15211 35 35 0.05 0.81598 1.60732
```

- JMP Data Table and formulas (Figure 3.4)

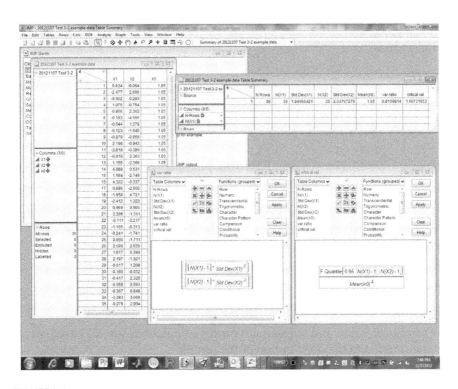

FIGURE 3.4
Test 3.2, JMP screen.

Test 3.3 Single Coefficient of Variation (One-Sided)

Parameters:

$$c = \frac{\sigma}{\mu} = \text{coefficient of variation (cv)}$$

c_0 = upper tolerable bound on c.

$1 - \beta$ = power to reject the null if c equals c_0.

Hypotheses:

$$H_0: c > c_0$$

$$H_1: c \le c_0$$

Data:

$$\hat{c} = \frac{\hat{\sigma}}{\hat{\mu}} = \text{sample cv.}$$

$$n = \text{sample size.}$$

Critical value(s):
   Reject H_0 if

$$\hat{c} \le \frac{\sqrt{n}}{T'\left(\beta, n-1; \dfrac{\sqrt{n}}{c_0}\right)}.$$

$$T'\left(\beta, n-1; \frac{\sqrt{n}}{c_0}\right)$$

is the $100(\beta)$ percentile of a noncentral t-distribution with $n - 1$ degrees of freedom and noncentrality parameter

$$\frac{\sqrt{n}}{c_0}.$$

Discussion:
   The critical value and subsequent power calculations are all based on an approximation to the distribution of the sample

$$\text{CV}, \hat{c} = \frac{\hat{\sigma}}{\hat{\mu}}.$$

Kang et al. (2007) used the approximation that

$$\hat{c} \sim \frac{\sqrt{n}}{T'\left(n-1, \frac{\sqrt{n}}{c}\right)}$$

where

$$T'\left(n-1, \frac{\sqrt{n}}{c}\right)$$

is a noncentral t random variable with $n - 1$ degrees of freedom and noncentrality parameter

$$\frac{\sqrt{n}}{c}.$$

Using T' to represent the appropriate noncentral t variable, the power to reject the null hypothesis is:

$$\Pr\left\{\hat{c} \le \frac{\sqrt{n}}{T'\left(\beta, n-1, \frac{\sqrt{n}}{c}\right)}\right\} = \Pr\left\{\frac{\sqrt{n}}{T'} \le \frac{\sqrt{n}}{T'\left(\beta, n-1, \frac{\sqrt{n}}{c}\right)}\right\}$$

$$= \Pr\left\{T' \ge T'\left(\beta, n-1, \frac{\sqrt{n}}{c}\right)\right\}.$$

Example:
 Suppose:

$$c_0 = 0.06$$

$$H_0: c > 0.06$$

$$H_1: c \le 0.06$$

$$\hat{c} = 0.065$$

$$n = 64$$

$$1 - \beta = 0.95$$

Then

$$\frac{\sqrt{n}}{c_0} = \frac{\sqrt{64}}{0.06} \approx 133.33$$

$$T'\left(\beta, n-1; \frac{\sqrt{n}}{c_0}\right) \approx 116.43$$

$$\frac{\sqrt{n}}{T'\left(\beta, n-1; \dfrac{\sqrt{n}}{c_0}\right)} \approx \frac{8}{116.43} \approx 0.0678.$$

Since $\hat{c} = 0.065 < 0.0687$, the null hypothesis is rejected. Figure 3.5 shows the power curve for this example.

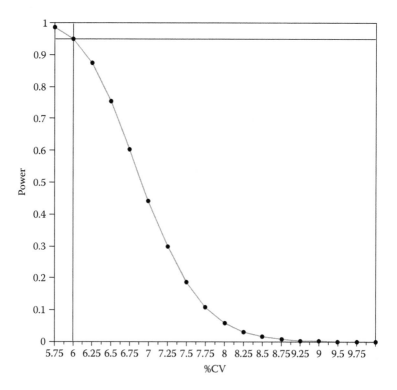

FIGURE 3.5
Test 3.3, power curve for test of single CV.

Confidence interval formulation:

$$\frac{\sqrt{n}}{T'\left(\beta, n-1; \frac{\sqrt{n}}{\hat{c}}\right)}$$

is a 100 percent $(1 - \beta)$ upper confidence limit on the coefficient of variation, c.

The example data yield:

$$\frac{\sqrt{n}}{T'\left(\beta, n-1; \frac{\sqrt{n}}{\hat{c}}\right)} = \frac{\sqrt{64}}{T'\left(0.05, 63; \frac{\sqrt{64}}{0.065}\right)} \approx \frac{8}{107.46} \approx 0.0744.$$

Computational considerations:

- SAS code

```
libname stuff 'H:\Personal Data\Equivalence & Noninferiority\
Programs & Output';

data calc;
   set stuff.d20121107_test_3_3_example_data;

   run;

proc means data = calc;
   var X c0 beta;
   output out = onemean MEAN = xbar c0val betaprob STD = sdx
   N = n1;
   run;

data outcalc;
   set onemean;
   samp_cv = sdx/xbar;
   crit_val = sqrt(n1)/tinv(betaprob,n1-1,sqrt(n1)/c0val);
   run;

proc print data = outcalc;/* has vars xbar c0val betaprob sdx
n1 sampe_cv crit_val */
   run;
```

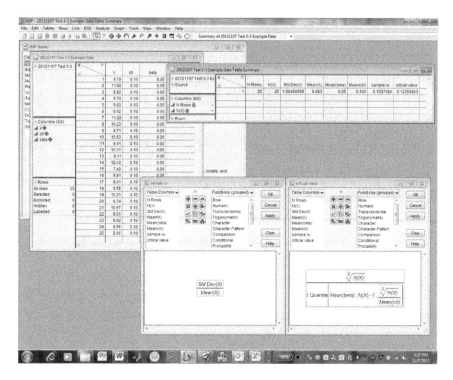

FIGURE 3.6
Test 3.3, JMP screen.

```
                        The MEANS Procedure

Variable   Label    N        Mean      Std Dev     Minimum      Maximum
ƒƒƒƒƒƒƒƒƒƒƒƒƒƒƒƒƒƒƒƒƒƒƒƒƒƒƒƒƒƒƒƒƒƒƒƒƒƒƒƒƒƒƒƒƒƒƒƒƒƒƒƒƒƒƒƒƒƒƒƒƒƒƒƒƒƒƒƒƒƒƒƒƒƒƒƒ
X          X        25   9.6829027   1.0045857   7.4892100   12.0170228
c0         c0       25   0.1000000           0   0.1000000    0.1000000
beta       beta     25   0.0500000           0   0.0500000    0.0500000
ƒƒƒƒƒƒƒƒƒƒƒƒƒƒƒƒƒƒƒƒƒƒƒƒƒƒƒƒƒƒƒƒƒƒƒƒƒƒƒƒƒƒƒƒƒƒƒƒƒƒƒƒƒƒƒƒƒƒƒƒƒƒƒƒƒƒƒƒƒƒƒƒƒƒƒƒ

        The SAS System 13:53 Wednesday, November 7, 2012 10

Obs _TYPE_  _FREQ_  xbar c0val betaprob sdx n1 samp_cv crit_val
 1     0       25   9.68  0.10    0.05  1.00 25 0.10375  0.12351
```

- JMP Data Table and formulas (Figure 3.6)

4

Exponential Rate Parameters

Test 4.1 Single Exponential Rate Parameter (One-Sided)

<u>Parameters:</u>
λ = exponential rate parameter (number of events per unit time)
λ_0 = maximum tolerable rate
$1 - \beta$ = power to reject the null if $\lambda = \lambda_0$

<u>Hypotheses:</u>

$$H_0: \lambda > \lambda_0$$

$$H_1: \lambda \le \lambda_0$$

<u>Data:</u>

$$\bar{T} = \text{average time between events}$$

$$\bar{\lambda} = \frac{1}{\bar{T}} = \text{average rate (number of events per unit time)}$$

<u>Critical value(s):</u>
If time to event, T, has an exponential distribution, then

$$\hat{\gamma} = \sum_{i=1}^{n} T_i = n\bar{T}$$

has a gamma distribution with parameters λ_0 and n. Often n is called the "shape" parameter, and λ_0 is called the "scale" parameter. Thus, if

$$G(\hat{\gamma} \mid n, \lambda_0) = \int_0^{\hat{\gamma}} \frac{\lambda_0^n}{\Gamma(n)} x^{n-1} e^{-\lambda_0 x} dx \ge \beta$$

then reject H_0. That is, $G(.)$ is the cumulative distribution function for a gamma-distributed random variable with parameters λ_0 and n. Similarly, find the value, call it γ_c, such that

$$G(\gamma_c \mid n, \lambda_0) = \int_0^{\gamma_c} \frac{\lambda_0^n}{\Gamma(n)} x^{n-1} e^{-\lambda_0 x} dx = \beta$$

and reject H_0 if

$$\hat{\gamma} \geq \gamma_c.$$

One could also rely on the central limit theorem, and reject the null hypothesis if

$$\bar{T} + t_{1-\beta} \frac{S}{\sqrt{n}} \geq \frac{1}{\lambda_0}$$

where S is the sample standard deviation of the T_i.

Discussion:
The smaller the rate parameter, the longer the expected time to event. Thus, demonstrating that the average time to event is at least $1/\lambda_0$ is the same as showing that the rate (number of events per fixed unit of time) is no greater than λ_0.

In this formulation, we have implicitly assumed that the "events" of interest are undesirable, such as a machine failure. Thus, high rates of occurrence would be undesirable.

The probability of rejecting the null hypothesis under the alternative that $\lambda = \lambda_a$ is

$$1 - G(\gamma_c \mid n, \lambda_a) = 1 - \int_0^{\gamma_c} \frac{\lambda_a^n}{\Gamma(n)} x^{n-1} e^{-\lambda_a x} dx.$$

Example:
Let $\lambda_0 = 0.75$, or $1/\lambda_0 \approx 1.33$. So the hypotheses are

$$H_0: \lambda > 0.75$$

$$H_1: \lambda \leq 0.75$$

Suppose:

$$\bar{T} = 1.29 \text{ (sample average time to event)}$$

$$S = 1.08$$

$$n = 20$$

$$\beta = 0.05$$

$$t_{1-\beta} \approx 1.729$$

$$\bar{T} + t_{1-\beta}\frac{S}{\sqrt{n}} \approx 1.29 + 1.729\frac{1.08}{\sqrt{20}} \approx 1.708 \geq \frac{1}{\lambda_0} \approx 1.33.$$

Therefore, we reject H_0. Also, we have

$$\hat{\gamma} \approx 20 * 1.29 = 25.8$$

and

$$1 - G(25.8 \,|\, n = 20, \lambda_0 = 0.75) = 1 - \int_0^{25.8} \frac{\lambda_0^n}{\Gamma(n)} x^{n-1} e^{-\lambda_0 x} dx \approx 0.5287 > \beta.$$

Therefore, we would reject H_0 using the "exact" gamma critical value. Note that for $\beta = 0.05$, with $n = 20$, the critical value, γ_c, is 17.67. That is,

$$G(17.67 \,|\, n = 20, \lambda_0 = 0.75) = \int_0^{17.67} \frac{\lambda_0^n}{\Gamma(n)} x^{n-1} e^{-\lambda_0 x} dx \approx 0.05 = \beta.$$

Figure 4.1 shows the power curve using the exact gamma critical value of 17.67 for this example.

Confidence interval formulation:

$$\bar{T} + t_{1-\beta}\frac{S}{\sqrt{n}}$$

is a $100(1 - \beta)$ percent upper confidence limit on $1/\lambda$. For the example,

$$\bar{T} + t_{1-\beta}\frac{S}{\sqrt{n}} \approx 1.29 + 1.729\frac{1.08}{\sqrt{20}} \approx 1.708$$

is a 95 percent upper confidence limit on $1/\lambda$, the mean time to event.

Computational considerations:
   Care should be taken when using built-in functions for computing probabilities or quantiles from gamma distributions. In some cases, the "scale" parameter may be expressed as a rate, for example, the expected number of occurrences per unit time, and in some cases, the scale parameter may be expressed as a mean time between events (R, JMP). In some cases,

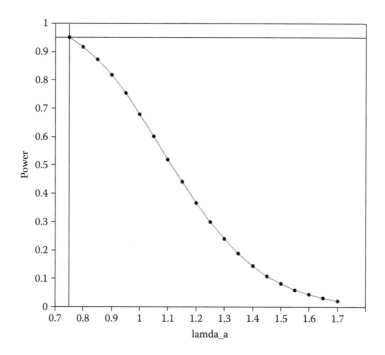

FIGURE 4.1
Test 4.1, power curve for exponential time-to-event test—exact gamma method.

such as SAS, the scale parameter is assumed to be 1, so that a change of variables is required to compute probabilities or quantiles from distributions where the scale parameter is not 1.

- SAS code

```
libname stuff 'H:\Personal Data\Equivalence & Noninferiority\
Programs & Output';

data calc;

  set stuff.d20121108_test_4_1_example_data;

  run;

proc means data = calc;
  var T lamda0 beta;
  output out = onemean MEAN = tbar lamda0val betaprob
  SUM = sumt STD = sdt N = n1;
  run;

data outcalc;
  set onemean;
```

```
crit_gam = gaminv(betaprob,n1)/lamda0val;
norm_stat = tbar + tinv(1-betaprob,n1-1)*sdt/sqrt(n1);
crit_norm = 1/lamda0val;
run;

proc print data = outcalc;/* has tbar lamda0val betaprob sumt
sdt n1 crit_gam norm_stat crit_norm */
  run;
```

```
                   The MEANS Procedure

Variable   Label     N      Mean     Std Dev   Minimum    Maximum
ƒƒƒƒƒƒƒƒƒƒƒƒƒƒƒƒƒƒƒƒƒƒƒƒƒƒƒƒƒƒƒƒƒƒƒƒƒƒƒƒƒƒƒƒƒƒƒƒƒƒƒƒƒƒƒƒƒƒƒƒƒƒƒ
T          T        25  1.9768811  1.6434899  0.0482092  5.6483261
lamda0     lamda0   25  1.1000000          0  1.1000000  1.1000000
beta       beta     25  0.0500000          0  0.0500000  0.0500000
ƒƒƒƒƒƒƒƒƒƒƒƒƒƒƒƒƒƒƒƒƒƒƒƒƒƒƒƒƒƒƒƒƒƒƒƒƒƒƒƒƒƒƒƒƒƒƒƒƒƒƒƒƒƒƒƒƒƒƒƒƒƒƒ
```

```
         The SAS System 08:52 Thursday, November 8, 2012 12

                                                     norm_   crit_
Obs _TYPE_  _FREQ_  tbar lamda0val betaprob  sumt     sdt  n1 crit_gam stat    norm
1     0       25  1.97688    1.1      0.05  49.42201.64349 25  15.8019 2.539240.90909
```

- JMP Data Table and formulas (Figure 4.2)

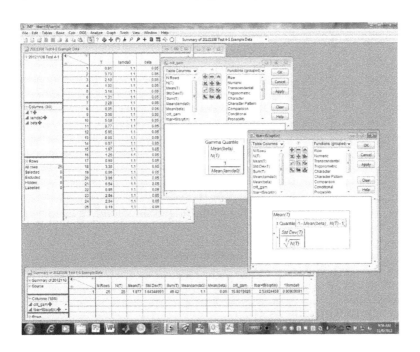

FIGURE 4.2
Test 4.1, JMP screen.

Test 4.2 Two Exponential Rate Parameters (One-Sided)

Parameters:

λ_c = exponential rate parameter (number of events per unit time), comparator system

λ_e = exponential rate parameter (number of events per unit time), evaluation system

n = number of times to event, comparator system (no censoring)

m = number of times to event, evaluation system (no censoring)

δ_0 = proportionality constant, $\delta_0 > 1$

T_c = time-to-event variable, comparator system

T_e = time-to-event variable, evaluation system

$E[T_c]$ = expected value of $T_c = 1/\lambda_c$

$E[T_e]$ = expected value of $T_e = 1/\lambda_e$

Hypotheses:

$$H_0: \lambda_e > \delta_0 \lambda_c$$

$$H_1: \lambda_e \leq \delta_0 \lambda_c$$

Or, alternatively:

$$H_0: E[T_e] < \frac{1}{\delta_0} E[T_c]$$

$$H_1: E[T_e] \geq \frac{1}{\delta_0} E[T_c]$$

Data:

$$\bar{T}_c = \text{mean time to event, comparator system}$$

$$\bar{T}_e = \text{mean time to event, evaluation system}$$

Critical value(s):

Reject the null hypothesis, H_0, if

$$\bar{T}_e - \frac{1}{\delta_0}\bar{T}_c + t_\beta SE \geq 0$$

where

$$SE = \sqrt{\frac{\overline{T}_e^2}{m} + \frac{\overline{T}_c^2}{n\delta_0^2}}$$

and $t_\beta = 100(1 - \beta)$ percentile of a (central) t-distribution with $m + n - 2$ degrees of freedom.

Discussion:

This test is a special case of Test 2.3, with the primary difference being the estimate of the standard error, SE. Under the assumption that T has an exponential distribution,

$$\hat{\sigma}^2[\overline{T}] = \frac{1/\hat{\lambda}^2}{n} = \frac{\overline{T}^2}{n}$$

is the minimum variance estimate for the variance of $\overline{T}$ (Mann, Schafer, and Singpurwalla, 1974).

Suppose $H_0: \lambda_e = \delta_a \lambda_c$, $\delta_a > \delta_0$. Then the test statistic

$$\frac{\overline{T}_e - \frac{1}{\delta_0} \overline{T}_c}{SE}$$

has a noncentral t-distribution with noncentrality parameter:

$$nc = \frac{\left(\dfrac{1}{\delta_a} - \dfrac{1}{\delta_0}\right)\sqrt{n}}{\sqrt{1 + \left(\dfrac{1}{\delta_0}\right)^2}}.$$

The correspondence between Test 4.2 and Test 2.3 is

$$\frac{1}{\delta_0} = 1 - p_0$$

The exponential distribution assumption simplifies the noncentrality, since in this case

$$E(T) = \mu = \sigma = \frac{1}{\lambda}.$$

Example:
  Suppose:

$\bar{T}_c = 100$
$\bar{T}_e = 80$
$n = m = 100$
$\delta = 1.001$

so that

$$SE = \sqrt{\frac{\bar{T}_e^2}{m} + \frac{\bar{T}_c^2}{n\delta^2}} = \sqrt{\frac{(80)^2}{100} + \frac{1}{1.001^2}\frac{(100)^2}{100}} \approx 12.798.$$

Then the test statistic is

$$\bar{T}_e - \frac{1}{\delta}\bar{T}_c + t_\beta SE \approx 80 - \frac{1}{1.0001}(100) + 1.653(12.798) \approx 1.250 \geq 0.$$

Therefore, the null hypothesis, H_0, is rejected.

Confidence interval formulation:

$$\bar{T}_e - \frac{1}{\delta}\bar{T}_c + t_\beta SE$$

is an approximate $100(1 - \beta)$ percent upper confidence limit on

$$E[T_e] - \frac{1}{\delta}E[T_c].$$

Computational considerations:
  See the code for Test 2.3.

5

Capability Indices

Test 5.1 C_{pk}

<u>Parameters:</u>
L = lower specification limit
U = upper specification limit
μ = population mean
σ = population standard deviation
n = sample size

$$C_{pL} = \frac{\mu - L}{3\sigma}$$

$$C_{pU} = \frac{U - \mu}{3\sigma}$$

$$C_{pk} = \min\left(C_{pL}, C_{pU}\right)$$

K_0 = minimum desirable value for C_{pk}

<u>Hypotheses:</u>

$$H_0: C_{pk} < K_0$$

$$H_1: C_{pk} \geq K_0$$

<u>Data:</u>

$$\bar{X} = \text{sample mean}$$

$$S = \text{sample standard deviation}$$

$$\hat{C}_{pL} = \frac{\bar{X} - L}{3S}$$

$$\hat{C}_{pu} = \frac{U - \bar{X}}{3S}$$

$$\hat{C}_{pk} = \min\left(\hat{C}_{pL}, \hat{C}_{pu}\right)$$

Critical value(s):

Both $\hat{C}_{pL}$ and $\hat{C}_{pu}$ have distributions that are proportional to a noncentral T. Specifically,

$$3\sqrt{n}\hat{C}_{pL} \sim T'\left(n-1, \delta = 3\sqrt{n}C_{pL}\right).$$

Thus, to test the hypotheses:

$$H_0: C_{pL} < K_0$$

$$H_1: C_{pL} \geq K_0$$

reject the null hypothesis if

$$\hat{C}_{pL} \geq \frac{T_\beta'\left(n-1, \delta = 3\sqrt{n}K_0\right)}{3\sqrt{n}}$$

where

$$T_\beta'\left(n-1, \delta = 3\sqrt{n}K_0\right)$$

is the lower 100βth percentile of a noncentral t-distribution with $n-1$ degrees of freedom and noncentrality parameter $\delta = 3\sqrt{n}K_0$. The same logic is used for $\hat{C}_{pu}$; namely, reject the null hypothesis if

$$\hat{C}_{pu} \geq \frac{T_\beta'\left(n-1, \delta = 3\sqrt{n}K_0\right)}{3\sqrt{n}}.$$

Note that since $\hat{C}_{pk} = \min\left(\hat{C}_{pL}, \hat{C}_{pu}\right)$, the decision to accept or reject is made based on only one of $\hat{C}_{pL}$ or $\hat{C}_{pu}$. The rejection probability should actually be based on the distribution of the minimum of two noncentral T random variables. However, for practical purposes, the fact that the test statistic is actually a function of two random variables will be ignored.

An approximation to the standard error of $\hat{C}_{pk}$ is described by Bissell (1990):

$$SE = \sqrt{\frac{1}{9n} + \frac{\hat{C}_{pk}^2}{2(n-1)}}.$$

As an alternative approach, a normal approximation could be used with this expression for the standard error. The critical value would be similar to the case of the single mean, namely:

Reject H_0 if:

$$\hat{C}_{pk} + z_{1-\beta}SE \geq K_0$$

where $z_{1-\beta} = 100*(1 - \beta)$ percentile of a standard normal distribution.

Discussion:
As in the case of all the hypothesis tests described in this work, we assume that the "null" value stated for C_{pk} is acceptable, so that we desire a high probability of rejecting the null hypothesis if the true C_{pk} is equal to the null value. Thus, the critical value is the lower 100βth percentile of the appropriate sampling distribution.

It should be noted that the estimates $\hat{C}_{pL}$ and $\hat{C}_{pu}$ are biased, in that their expected values exceed the value of the parameter. The bias is a multiplier described by:

$$bias = \sqrt{\frac{n-1}{2}} \frac{\Gamma\left(\dfrac{n-2}{2}\right)}{\Gamma\left(\dfrac{n-1}{2}\right)}$$

where $\Gamma(.)$ is the gamma function. The bias is relatively small for values of $n \geq 100$. At $n = 100$, the bias is approximately 1.008, or 0.8 percent too high. To compensate for bias, multiply the critical value by the *bias* formula. Alternatively, divide the usual estimator by the bias, and then use the "unadjusted" critical value.

The power to reject the null is given by:

$$\Pr\left\{\hat{C}_{pL} \geq \frac{T_{\beta}'\left(n-1, \delta = 3\sqrt{n}K_0\right)}{3\sqrt{n}} \mid \delta_a = 3\sqrt{n}K_a\right\}$$

and

$$T_{\beta}'\left(n-1, \delta = 3\sqrt{n}K_0\right)$$

is the lower 100βth percentile of a noncentral t-distribution with $n-1$ degrees of freedom and noncentrality parameter $\delta = 3\sqrt{n}K_0$. K_a is the alternate value of the population C_{pL}.

Note that the test for C_{pk} is identical in theory to the criteria used in ANSI ASQ Z1.9 standard sampling plans (1993), or ISO 3951 (1989). That is, the statistics:

$$\hat{C}_{pL} = \frac{\overline{X} - L}{3S}$$

and

$$\hat{C}_{pU} = \frac{U - \overline{X}}{3S}$$

are replaced in the standard sampling plans by

$$K_L = \frac{\overline{X} - L}{S}$$

and

$$K_U = \frac{U - \overline{X}}{S}$$

which only differ from the C_{pk} statistics by the multiplicative constant 3. Thus, power characteristics for the hypothesis tests concerning C_{pk} are identical to those of the ANSI/ASQ Z1.9 or ISO 3951 plans.

Example:
    Suppose that K_0 = 1.33. With a sample of n = 100, suppose that $\hat{C}_{pL} = 1.30$ and $\hat{C}_{pU} = 2.00$. Then the decision to reject the null hypothesis will be based entirely on $\hat{C}_{pL}$. For β = 0.05, the noncentrality is

$$\delta = 3\sqrt{n}K_0 = 39.9$$

and the critical value for $\hat{C}_{pL} \approx 1.18$. Since $\hat{C}_{pL} = 1.30$, the null hypothesis is rejected. The bias-corrected critical value is approximately 1.19. Figure 5.1 shows a power curve for this test.

Confidence interval formulation:
    For the exact lower 100(1 − β) percent confidence limit on C_{pL} or C_{pU} find $\hat{C}_{pL}(low)$ such that

$$\Pr\left\{T'\left(n-1, \delta = 3\sqrt{n}\hat{C}_{pL}(low)\right) \leq \hat{C}_{pL}\right\} = 1 - \beta$$

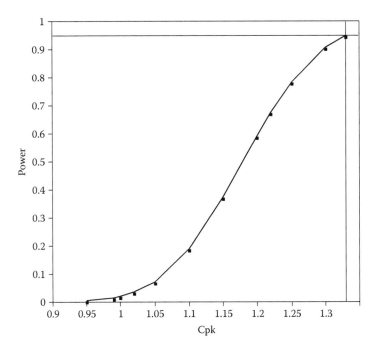

FIGURE 5.1
Test 5.1, power curve for noninferiority test on C_{pk}, $K_0 = 1.33$, $n = 100$.

or $\hat{C}_{pu}(low)$ such that

$$\Pr\left\{T'\left(n-1, \delta = 3\sqrt{n}\hat{C}_{pu}(low)\right) \le \hat{C}_{pu}\right\} = 1 - \beta.$$

See Kushler and Hurley (1992).
   Using the normal approximation, the lower confidence limit would be:

$$\hat{C}_{pL} - z_{1-\beta}\sqrt{\frac{1}{9n} + \frac{\hat{C}_{pL}^2}{2(n-1)}} \quad \text{or}$$

$$\hat{C}_{pu} - z_{1-\beta}\sqrt{\frac{1}{9n} + \frac{\hat{C}_{pu}^2}{2(n-1)}}$$

Computational considerations:

- SAS code

```
libname stuff 'H:\Personal Data\Equivalence & Noninferiority\
Programs & Output';

data calc;

  set stuff.d20121108_test_5_1_example_data;

  run;

proc means data=calc;
  var X K0 beta LL;
  output out = onemean MEAN = xbar K0val betaprob LLval
STD = sdx N = nx;
  run;

data outcalc;

  set onemean;
  CpL = (xbar - LLval)/(3*sdx);
  bias = sqrt((nx-1)/2)*gamma((nx-2)/2)/gamma((nx-1)/2);
  CpLcorr = CpL/bias;
  crit_val = tinv(betaprob,nx-1,3*sqrt(nx)*K0val)/
(3*sqrt(nx));
  run;

proc print data=outcalc;/* has vars xbar K0val betaprob LLval
sdx nx CpL bias CpLcorr crit_val */

  run;
```

The SAS System 08:52 Thursday, November 8, 2012 19

The MEANS Procedure

Variable	Label	N	Mean	Std Dev	Minimum	Maximum
X	X	25	99.6029607	1.8757915	95.5723262	102.2447737
K0	K0	25	0.7200000	0	0.7200000	0.7200000
beta	beta	25	0.0500000	0	0.0500000	0.0500000
LL	LL	25	95.0000000	0	95.0000000	95.0000000

The SAS System 08:52 Thursday, November 8, 2012 20

Obs	_TYPE_	_FREQ_	xbar	K0val	betaprob	LLval	sdx	nx	CpL	bias	CpLcorr	crit_val
1	0	25	99.60	0.72	0.05	95	1.88	25	0.81796	1.03267	0.79208	0.55321

- JMP Data Table and formulas (Figures 5.2 and 5.3)

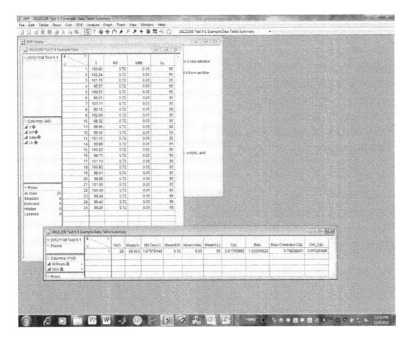

FIGURE 5.2
Test 5.1, JMP screen 1.

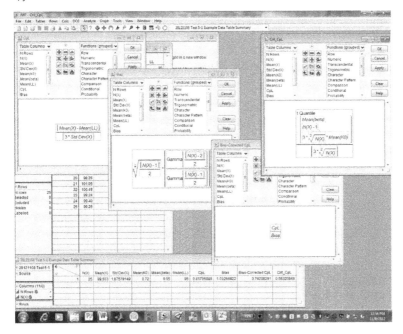

FIGURE 5.3
Test 5.1, JMP screen 2.

Test 5.2 C_p

Parameters:
  L = lower specification limit
  U = upper specification limit
  μ = population mean
  σ = population standard deviation
  n = sample size

$$C_p = \frac{U - L}{6\sigma}$$

  K_0 = minimum desirable value for C_p

Hypotheses:

$$H_0: C_p < K_0$$

$$H_1: C_p \geq K_0$$

$$K_0 = \frac{U - L}{6\sigma_0} \Rightarrow \sigma_0 = \frac{U - L}{6K_0}$$

so the hypotheses are equivalent to

$$H_0: \sigma^2 > \sigma_0^2 = \left(\frac{U - L}{6K_0}\right)^2$$

$$H_1: \sigma^2 \leq \sigma_0^2 = \left(\frac{U - L}{6K_0}\right)^2$$

Data:

$$S = \text{sample standard deviation}$$

$$\hat{C}_p = \frac{U - L}{6S}$$

Critical value(s):
  The critical value for S is:

$$S^2 \leq \frac{\sigma_0^2}{n-1}\chi_{1-\beta}^2(n-1) = \left(\frac{U - L}{6K_0(n-1)}\right)^2 \chi_{1-\beta}^2(n-1)$$

where $\chi^2_{1-\beta}(n-1)=100(1-\beta)$ percentile of a chi-squared distribution with $n - 1$ degrees of freedom. Therefore, the critical value for

$$\hat{C}_p = \frac{U-L}{6S}$$

is

$$K_c = \frac{\sqrt{(n-1)}K_0}{\sqrt{\chi^2_{1-\beta}(n-1)}}$$

where $\chi^2_{1-\beta}(n-1)=100(1-\beta)$ percentile of a chi-squared distribution with $n - 1$ degrees of freedom.

Discussion:
   Power calculations are very similar to those for the single standard deviation case. Simply make the substitution:

$$\sigma_0 = \frac{U-L}{6K_0}.$$

Example:
   Suppose

$$K_0 = \frac{U-L}{6\sigma_0} = 1.33.$$

Suppose further that $n = 26$, and

$$\hat{C}_p = \frac{U-L}{6S} = 1.20.$$

The critical value, with $\beta = 0.05$, is

$$K_c = \frac{\sqrt{(n-1)}K_0}{\sqrt{\chi^2_{1-\beta}(n-1)}} \approx \frac{\sqrt{25}(1.33)}{\sqrt{37.65}} \approx 1.084.$$

Since

$$\hat{C}_p = \frac{U-L}{6S} = 1.20 \geq K_c \approx 1.084$$

reject H_0.

<u>Confidence interval formulation:</u>

The lower $100(1 - \beta)$ percent confidence limit for C_p is

$$C_p(L) = \frac{\sqrt{(n-1)}\hat{C}_p}{\sqrt{\chi^2_{1-\beta}(n-1)}}.$$

The upper $100(1 - \beta)$ percent confidence limit for C_p is

$$C_p(U) = \frac{\sqrt{(n-1)}\hat{C}_p}{\sqrt{\chi^2_{\beta}(n-1)}}.$$

<u>Computational considerations:</u>

- SAS code

```
libname stuff 'H:\Personal Data\Equivalence & Noninferiority\
Programs & Output';

data calc;

   set stuff.d20121108_test_5_2_example_data;

   run;

proc means data = calc;
   var X K0 beta L U;
   output out = onemean MEAN = xbar K0val betaprob Lval Uval
   STD = sdx N = nx;
   run;

data outcalc;

   set onemean;
   Cp = (Uval - Lval)/(6*sdx);
   crit_val = sqrt(nx-1)*K0val/sqrt(cinv(1-betaprob,nx-1));
   run;

proc print data = outcalc;/* has vars xbar K0val betaprob Lval
Uval sdx nx Cp crit_val */

   run;
```

The SAS System 08:52 Thursday, November 8, 2012 21

The MEANS Procedure

Variable	Label	N	Mean	Std Dev	Minimum	Maximum
ffffffffffffffffffffffffffffffffffffffffffffffffffffffffffffffffffffffff
X	X	30	99.9869696	0.7676268	97.5063021	101.2103010
K0	K0	30	0.8600000	0	0.8600000	0.8600000
beta	beta	30	0.0500000	0	0.0500000	0.0500000
L	L	30	98.0000000	0	98.0000000	98.0000000
U	U	30	102.0000000	0	102.0000000	102.0000000
ffffffffffffffffffffffffffffffffffffffffffffffffffffffffffffffffffffffff

The SAS System 08:52 Thursday, November 8, 2012 22

Obs	_TYPE_	_FREQ_	xbar	K0val	betaprob	Lval	Uval	sdx	nx	Cp	crit_val
1	0	30	99.99	0.86	0.05	98	102	0.77	30	0.86848	0.70992

- JMP Data Table and formulas (Figure 5.4)

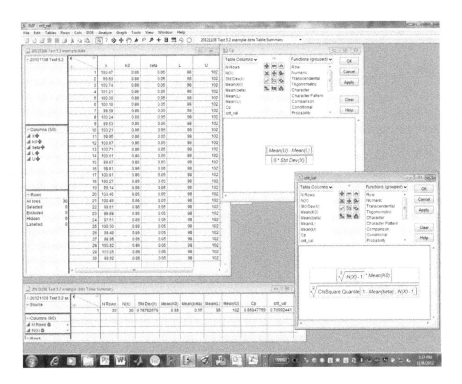

FIGURE 5.4
Test 5.2, JMP screen.

6

Multivariate

Test 6.1 Multivariate—Single Mean Vector (Two-Sided)

<u>Parameters:</u>

k = number of univariate variables being considered simultaneously

$\underline{\mu}$ = mean vector ($k \times 1$)

Σ = covariance matrix ($k \times k$)

$\underline{\mu_0}$ = target mean vector ($k \times 1$)

$\underline{\Delta}$ = tolerable distance between μ and μ_0, that is, noninferiority is defined as:

$$\left\|\underline{\mu} - \underline{\mu_0}\right\| \leq \Delta$$

where:

$$\|\delta\| = \sqrt{d_1^2 + d_2^2 + \cdots + d_k^2}$$

where:

$$\delta = \begin{bmatrix} d_1 \\ d_2 \\ \vdots \\ d_k \end{bmatrix}$$

is a $k \times 1$ column vector. $1 - \beta$ = power to reject the null if

$$\left\|\underline{\mu} - \underline{\mu_0}\right\| > \Delta.$$

<u>Hypotheses:</u>

$$H_0 : \left\|\underline{\mu} - \underline{\mu_0}\right\| > \Delta$$

$$H_1 : \left\|\underline{\mu} - \underline{\mu_0}\right\| \leq \Delta$$

Data:

X = (n × k) matrix of observations, x_{ij} = the *i*th observation of the *j*th variable.

$$
\hat{\mu} = \begin{bmatrix} \bar{x}_1 \\ \bar{x}_2 \\ \vdots \\ \bar{x}_k \end{bmatrix} = \text{the sample mean vector.}
$$

$$\hat{\Sigma} = \text{the sample covariance matrix.}$$

Critical value(s):
  Let

$$
T^2 = n\left(\hat{\mu} - \mu_0\right)^T \hat{\Sigma}^{-1}\left(\hat{\mu} - \mu_0\right).
$$

Then reject H_0 if:

$$
F' = \frac{n-k}{(n-1)k}T^2 \le F'_{1-\beta}(k, n-k, n\delta^T\hat{\Sigma}^{-1}\delta)
$$

where

$$
F'_{1-\beta}(k, n-k, n\delta^T\hat{\Sigma}^{-1}\delta)
$$

is the 100(1 − β) percentile of a noncentral *F*-distribution with degrees of freedom *k* (numerator) and *n − k* (denominator) and noncentrality parameter $n\delta^T\hat{\Sigma}^{-1}\delta$.

Discussion:
  The statistic:

$$
T^2 = n\left(\hat{\mu} - \mu_0\right)^T \hat{\Sigma}^{-1}\left(\hat{\mu} - \mu_0\right)
$$

has a distribution known as Hotelling's T^2. It can be shown (Anderson, 1958) that

$$
F' = \frac{n-k}{(n-1)k}T^2
$$

has a noncentral *F*-distribution with degrees of freedom *k* (numerator) and
$n - k$ (denominator); noncentrality parameter $n\delta^T \Sigma_0^{-1}\delta$. Σ_0 is the population
covariance matrix, which generally is unknown. Thus, the critical value for
the hypothesis test is based on the sample covariance matrix. Strictly speak-
ing, the critical value should be based in part on the population covariance
matrix. However, since it is generally unknown, substituting the sample
covariance matrix is a reasonable approximation.

Defining a multivariate noninferiority region presents some difficulties.
For one, it is possible that every dimension could satisfy the univariate
criteria:

$$\mu_{i0} - d_i \le \mu_i \le \mu_{i0} + d_i \quad \text{for } i = 1, k$$

and not satisfy the multivariate criterion:

$$\left\|\underline{\mu} - \underline{\mu_0}\right\| \le \Delta.$$

Example:
  Suppose the population mean vector is

$$\underline{\mu} = \begin{bmatrix} \mu_1 \\ \mu_2 \end{bmatrix} = \begin{bmatrix} 0.3 \\ 0.5 \end{bmatrix}$$

and

$$\underline{\mu_0} = \begin{bmatrix} \mu_{1,0} \\ \mu_{2,0} \end{bmatrix} = \begin{bmatrix} 0.4 \\ 0.4 \end{bmatrix}$$

so that:

$$\delta = \begin{bmatrix} d_1 \\ d_2 \end{bmatrix} = \begin{bmatrix} 0.1 \\ 0.1 \end{bmatrix}.$$

Furthermore, suppose that for each univariate mean, as well as in a
multivariate sense, the maximum tolerable difference was

$$\Delta = 0.125.$$

Then

$$\mu_{1,0} - d_1 = 0.4 - 0.1 = 0.3 \le (\mu_1 = 0.3) \le \mu_{1,0} + d_1 = 0.4 + 0.1 = 0.5$$

$$\mu_{2,0} - d_2 = 0.4 - 0.1 = 0.3 \le (\mu_2 = 0.5) \le \mu_{2,0} + d_2 = 0.4 + 0.1 = 0.5.$$

However,

$$\left\|\underline{\mu} - \underline{\mu}_0\right\| = \sqrt{(\mu_1 - \mu_{1,0})^2 + (\mu_2 - \mu_{2,0})^2} \approx 0.14142 > \Delta = 0.125$$

Thus, each univariate mean satisfies the univariate definitions for noninferiority, but the multivariate criterion is not satisfied.
Confidence interval formulation:

In a multivariate situation, the confidence region is a k-dimensional object and its interior. In the case of mean vectors, it is an ellipsoid together with its interior. Specifically, it is the set of all vectors, μ, such that

$$T^2 = n\left(\hat{\underline{\mu}} - \underline{\mu}\right)^T \hat{\Sigma}^{-1}\left(\hat{\underline{\mu}} - \underline{\mu}\right) \le \frac{k(n-1)}{n-k} F_{\alpha,k,n-k}$$

where $F_{\alpha, k, n-k}$ = the $100(1 - \alpha)$ percentile of a (central) F-distribution with k numerator degrees of freedom and $n - k$ denominator degrees of freedom, and

$$\hat{\underline{\mu}} = \begin{bmatrix} \bar{x}_1 \\ \bar{x}_2 \\ \vdots \\ \bar{x}_k \end{bmatrix} = \text{the sample mean vector.}$$

Computational considerations:

While it is possible to use JMP scripting language (JSL) or SAS Proc IML to compute and invert covariance matrices, it is easier to do so in R. Use the R function cov() to compute the covariance matrix, and the solve() function to invert the covariance matrix. Recall that in R, a statement of the form:

```
>x <- c(1, 2, 3)
```

creates a column vector, not a row vector, called x.

R:

```
> df1 <- read.table("H:\\Personal Data\\Equivalence &
Noninferiority\\Programs & Output\\d20121109_test_6_1_example.
csv",header = TRUE,sep = ",")
> attach(df1)
> xmat <- as.matrix(df1)
> xmat
         X1       X2       X3
[1,]  100.21   33.37   102.22
[2,]  101.22   33.79   100.33
[3,]   97.16   32.32    95.55
[4,]   98.72   33.05    97.60
```

```
 [5,]     97.26    32.41     97.01
 [6,]    100.71    33.34    101.12
 [7,]    101.30    33.75    101.62
 [8,]     98.88    32.99     97.98
 [9,]    100.25    33.55     99.88
[10,]     97.19    32.19     96.81
[11,]    105.16    35.07    105.63
[12,]     98.30    32.70     96.40
[13,]    100.74    33.38    101.39
[14,]     97.69    32.62     98.26
[15,]    100.02    33.12    101.32
[16,]    101.43    33.94    102.32
[17,]     95.40    31.76     96.44
[18,]     99.85    33.08     99.45
[19,]     97.43    32.43     98.82
[20,]    100.63    33.59     99.84
[21,]    102.00    33.90    101.51
[22,]     97.45    32.30     97.14
[23,]    102.03    34.03    100.69
[24,]     99.79    33.43     99.40
[25,]     98.61    33.00     97.65
[26,]     95.31    31.77     93.97
[27,]     98.34    32.57     99.32
[28,]     97.27    32.49     96.80
[29,]     98.46    32.82     99.50
[30,]    100.42    33.50     98.54
[31,]     99.90    33.39     99.34
[32,]    101.10    33.59     99.98
[33,]     98.40    32.72     96.44
[34,]    100.80    33.73    101.81
[35,]     95.56    31.62     94.11
[36,]     98.63    32.72    100.50
[37,]     99.13    32.80     98.35
[38,]    103.18    34.52    102.33
[39,]    101.08    33.64    100.17
[40,]    102.08    33.93    103.59
> cmat <- cov(xmat)
> cmat
        X1          X2          X3
X1 4.578826 1.5819854 4.853564
X2 1.581985 0.5612728 1.654349
X3 4.853564 1.6543490 6.324297
> cmatinv <- solve(cmat)
> cmatinv
         X1          X2          X3
X1  10.549891 -25.641264 -1.3890829
X2 -25.641264  70.101515  1.3407270
X3  -1.389083   1.340727  0.8734525
> mu1_est <- mean(X1)
> mu2_est <- mean(X2)
```

```
> mu3_est <- mean(X3)
> mu_est <- c(mu1_est,mu2_est,mu3_est)
> mu_est
[1] 99.47725 33.12300 99.27825
> mu0 <- c(100,32,99)
> mu0
[1] 100 32 99
> diff <- mu_est - mu0
> diff
[1] -0.52275 1.12300 0.27825
> n <- length(X1)
> n
[1] 40
> T2 <- n*t(diff)%*% cmatinv%*% diff
> T2
      [,1]
[1,] 4908.197
> k <- 3
> Fstat <- (n-k)*T2/((n-1)*k)
> Fstat
      [,1]
[1,] 1552.165
> delta <- c(3,2,3)
> nct <- n*t(delta)%*%cmatinv%*%delta
> nct
      [,1]
[1,] 2664.248
> Fcrit <- qf(0.95,k,n-k,nct)
> Fcrit
[1] 1371.96
```

Since Fstat > Fcrit, we fail to reject H_0, concluding that the mean vector is not equivalent to μ_0.

The function t(X) in R returns the transpose of the matrix (or vector) X.

7

Reliability

Test 7.1 Comparison of Reliability Using Parametric Models

Parameters:

T = time to failure

$h_c(t) = \alpha_0 + 2\alpha_1 t$ = hazard rate function, comparator system

$h_e(t) = \beta_0 + 2\beta_1 t$ = hazard rate function, evaluation system

(these hazard rate functions are a slight generalization of an exponential hazard rate, where $\alpha_1 = \beta_1 = 0$)

$$H_c(t) = \int_0^t h_c(\tau)d\tau = \alpha_0 t + \alpha_1 t^2$$

= cumulative hazard function, comparator system

$$H_e(t) = \int_0^t h_e(\tau)d\tau = \beta_0 t + \beta_1 t^2$$

= cumulative hazard function, evaluation system

$R_c(t) = e^{-H_c(t)} = \Pr\{T \ge t \mid h_c\}$ = reliability function, comparator system

$R_e(t) = e^{-H_e(t)} = \Pr\{T \ge t \mid h_e\}$ = reliability function, evaluation system

T = time at which reliability of the evaluation system is to be compared to that of the comparator system

δ = maximum allowable reduction in reliability of evaluation system compared to the comparator system at time T

Hypotheses:

$$H_0: R_e(T) < (1-\delta)R_c(T)$$

$$H_1: R_e(T) \ge (1-\delta)R_c(T)$$

Alternatively, the hypotheses can be stated as:

$$H_0: -H_e(T) + H_c(T) < \ln(1-\delta)$$

$$H_1: -H_e(T) + H_c(T) \geq \ln(1-\delta).$$

Based on the parametric model, these hypotheses can then be expressed as:

$$H_0: (\alpha_0 - \beta_0)T + (\alpha_1 - \beta_1)T^2 < \ln(1-\delta)$$

$$H_1: (\alpha_0 - \beta_0)T + (\alpha_1 - \beta_1)T^2 \geq \ln(1-\delta)$$

Or

$$H_0: D_0 T + D_1 T^2 < \ln(1-\delta)$$

$$H_1: D_0 T + D_1 T^2 \geq \ln(1-\delta)$$

$$D_k = \alpha_k - \beta_k, \ k = 0, 1, \text{ or finally:}$$

$$H_0: D_1 + \frac{1}{T}D_0 - \frac{1}{T^2}\ln(1-\delta) < 0$$

$$H_1: D_1 + \frac{1}{T}D_0 - \frac{1}{T^2}\ln(1-\delta) \geq 0$$

Data:

$t_{c1}, t_{c2}, \ldots, t_{cn} = n$ sample order statistics for failure times, comparator system

$t_{e1}, t_{e2}, \ldots, t_{em} = m$ sample order statistics for failure times, evaluation system

$r(t_{c1}), r(t_{c2}), \ldots, r(t_{cn}) =$ sample reliabilities, comparator system

$r(t_{e1}), r(t_{e2}), \ldots, r(t_{em}) =$ sample reliabilities, evaluation system

$$r(t_k) = \left(\frac{n-1}{n}\right)^{\delta_1} \prod_{i=2}^{k} \left(\frac{(n-i-1)-1}{n-i-1}\right)^{\delta_i} \cdot \text{(Kaplan-Meier estimator, see Lee, 1992)}$$

$$\text{with } \delta_i = \begin{cases} 1 & i^{th} \text{ obs. uncensored} \\ 0 & i^{th} \text{ obs. censored} \end{cases}$$

$\hat{H}_c(t) = \hat{\alpha}_0 t + \hat{\alpha}_1 t^2 =$ least squares fit of cumulative hazard, comparator system, with response variable $y = -\ln r(t_i)$.

$\hat{H}_e(t) = \hat{\beta}_0 t + \hat{\beta}_1 t^2 =$ least squares fit of cumulative hazard, evaluation system.

$$\hat{D}_k = \hat{\alpha}_k - \hat{\beta}_k, \ k = 0,1.$$

Critical value(s):
  Reject H_0 if

$$\hat{D}_1 + \frac{1}{T}\hat{D}_0 - \frac{1}{T^2}\ln(1-\delta) + t_\beta SE_{reg} \geq 0$$

where $t_\beta = 100(1 - \beta)$ percentile of a (central) t-distribution with $m + n - 4$ degrees of freedom, and

$$SE_{reg} = \sqrt{SE^2(\hat{D}_1) + \frac{1}{T^2}SE^2(\hat{D}_0)}$$

where

$$SE^2(\hat{D}_k) = SE^2(\hat{\alpha}_k) + SE^2(\hat{\beta}_k) \ k = 0,1 \text{ and } SE(\hat{\alpha}_k), SE(\hat{\beta}_k)$$

are standard errors for the regression estimates of the corresponding parameters.

Discussion:
  The parametric models for the hazard functions,

$$h_c(t) = \alpha_0 + 2\alpha_1 t$$

and

$$h_e(t) = \beta_0 + 2\beta_1 t$$

could be somewhat generalized to a pth-order polynomial in t:

$$h(t) = \sum_{k=0}^{p}(k+1)\gamma_k t^k.$$

Thus, the cumulative hazard function would have the form:

$$H(t) = \int_0^t h(\tau)d\tau = \sum_{k=0}^{p} \gamma_k t^{k+1}.$$

Example:

Suppose from a sample of $m = n = 100$ times to failure for two systems, least squares estimates of the parameters are

$$\hat{\alpha}_0 = 0.60902$$

$$\hat{\alpha}_1 = 0.01696$$

$$\hat{\beta}_0 = 0.74436$$

$$\hat{\beta}_1 = 0.17178$$

and the standard errors are

$$SE(\hat{\alpha}_0) \approx 0.014285$$

$$SE(\hat{\alpha}_1) \approx 0.002053$$

$$SE(\hat{\beta}_0) \approx 0.008214$$

$$SE(\hat{\beta}_1) \approx 0.002445$$

Figure 7.1 shows the sample reliability curves, $R_c(t)$ and $R_e(t)$.
Thus,

$$\hat{D}_0 = \hat{\alpha}_0 - \hat{\beta}_0 = 0.60902 - 0.74436 \approx -0.13534$$

$$\hat{D}_1 = \hat{\alpha}_1 - \hat{\beta}_1 = 0.01696 - 0.17178 \approx -0.15482$$

At $T = 0.5$, with $\delta = 0.10$,

$$SE(\hat{D}_0) \approx 0.016478$$

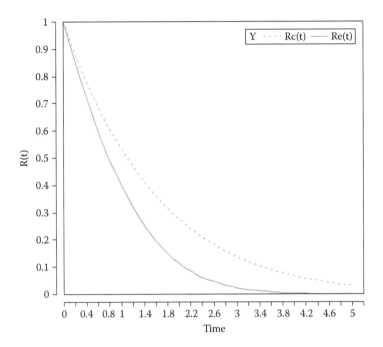

FIGURE 7.1
Sample reliability curves.

$$SE(\hat{D}_1) \approx 0.003193$$

$$SE_{reg} \approx 0.03311$$

With $100 + 100 - 4 = 196$ degrees of freedom, the statistic is

$$\hat{D}_1 + \frac{1}{T}\hat{D}_0 - \frac{1}{T^2}\ln(1-\delta) + t_\beta SE_{reg} \approx 0.000108 \geq 0.$$

Therefore, reject the null, and conclude that the evaluation system is equivalent to the comparator in reliability at $T = 0.5$.

Confidence interval formulation:
   The one-sided upper $100(1-\beta)$ percent confidence limit on $\hat{D}_k = \hat{\alpha}_k - \hat{\beta}_k$, $k = 0, 1$ is

$$\hat{D}_k + t_\beta SE(\hat{D}_k) = \hat{\alpha}_k - \hat{\beta}_k + t_\beta \sqrt{SE^2(\hat{\alpha}_k) + SE^2(\hat{\beta}_k)}$$

where $t_\beta = 100(1 - \beta)$ percentile of a (central) t-distribution with $m + n - 4$ degrees of freedom.

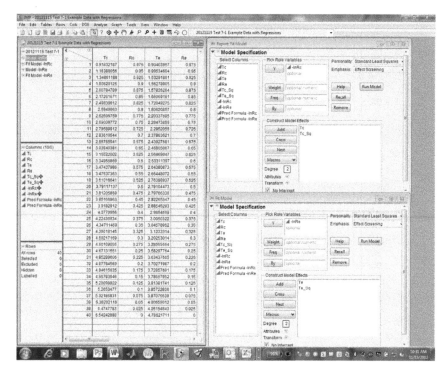

FIGURE 7.2
Test 7.1, JMP screen 1.

Computational considerations:

Computationally, Test 7.1 is a combination of Test 2.5 (regression slope) and Test 2.2 (comparison of two means, fixed Δ). In the example used here, the two systems are considered equivalent at time $T = 3$ if the reliability of the evaluation system, $Re(T)$, is at least 99 percent of $Rc(T)$ (i.e., $\delta = 0.01$).

JMP Data Table and formulas (Figure 7.2 and Figure 7.3). The columns re and rc in Figure 7.2 represent the reliability functions for the time-to-event variables, Te and Tc. You can use the "Fit Model" function to obtain coefficient estimates. Make sure to check the "No Intercept" option box.

Right-click on the Parameter Estimates Chapter of the Fit Model output, and select the "Make into Data Table" option. Create new columns to compute the standard error squared, and a column that indicates the row number in each parameter estimate table (called "order" in this example). Join the tables for Te and Tc parameter estimates using the Tables -> Join function, with "order" as the matching column. Then compute Dk, the difference in the parameter estimates, for the linear (order = 1) and quadratic (order = 2) terms of the two models.

- The models and the data (Figures 7.4 and 7.5):

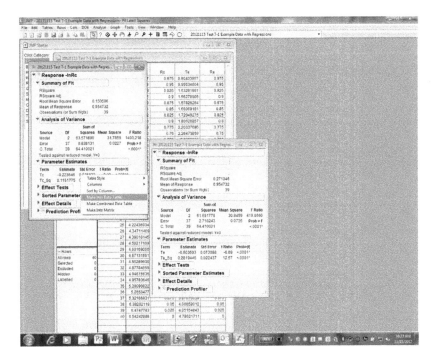

FIGURE 7.3
Test 7.1, JMP screen 2.

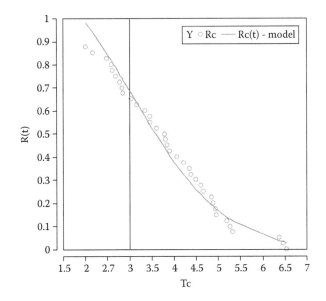

FIGURE 7.4
$R_c(t)$ (JMP overlay plot).

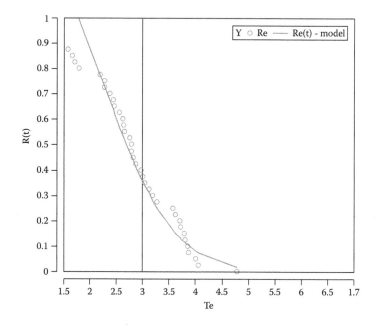

FIGURE 7.5

$R_e(t)$ (JMP overlay plot).

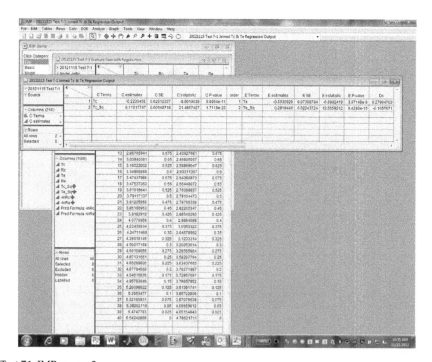

Test 7.1, JMP screen 3.

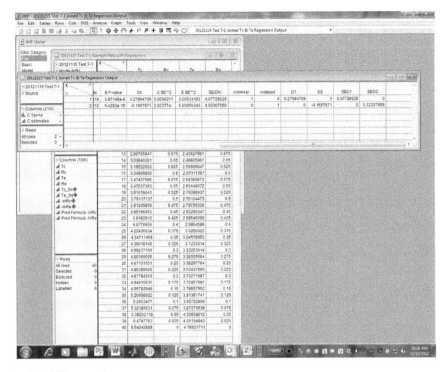

Test 7.1, JMP screen 4.

Create "dummy variable" columns ("indlinear" and "indquad") to allow the standard errors for the linear and quadratic coefficient estimates to be computed in the same row of an output table. That is, for example, indlinear = 1 for row 1 and 0 for row 2 of the estimate table. D1 is indlinear*Dk; D2 is indquad*Dk.

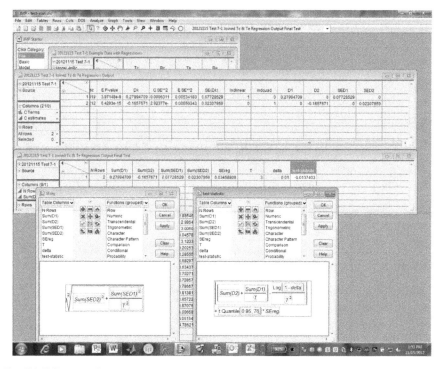

Test 7.1, JMP screen 5.

The test statistic requires as input a particular time to event (*T*) and δ. Since in this example the test statistic is less than 0, the null hypothesis is not rejected, and the conclusion is that the two systems are not equivalent at time *T* = 3.

Test 7.2 Nonparametric Test for Comparison of Reliability

Parameters:

    T = time to failure

    $h_c(t)$ = hazard rate function, comparator system (no form specified)

    $h_e(t)$ = hazard rate function, evaluation system (no form specified)

$$H_c(t) = \int_0^t h_c(\tau)d\tau = \text{cumulative hazard function, comparator system}$$

$$H_e(t) = \int_0^t h_e(\tau)d\tau = \text{cumulative hazard function, evaluation system}$$

$R_c(t) = e^{-H_c(t)} = \Pr\{T \geq t \mid h_c\}$ = reliability function, comparator system

$R_e(t) = e^{-H_e(t)} = \Pr\{T \geq t \mid h_e\}$ = reliability function, evaluation system

T = time at which reliability of the evaluation system is to be compared to that of the comparator system

δ = maximum allowable reduction in reliability of evaluation system compared to the comparator system at time T

Hypotheses:

$$H_0: R_e(T) < (1-\delta) R_c(T)$$

$$H_1: R_e(T) \geq (1-\delta) R_c(T)$$

Note that in this "nonparametric" case the hypotheses about $R(t)$ will be tested directly using estimates of $R(t)$.

Data:

$t_{c1}, t_{c2}, \ldots, t_{cn} = n$ sample order statistics for failure times, comparator system

$t_{e1}, t_{e2}, \ldots, t_{em} = m$ sample order statistics for failure times, evaluation system

$r(t_{c1}), r(t_{c2}), \ldots, r(t_{cn})$ = sample reliabilities, comparator system

$r(t_{e1}), r(t_{e2}), \ldots, r(t_{em})$ = sample reliabilities, evaluation system

And the Kaplan-Meier estimators:

$$\hat{R}_c(t_k) = \frac{n-1}{n} \prod_{i=2}^{k} \frac{(n-i-1)-1}{n-i-1}$$

$$\hat{R}_e(t_k) = \frac{m-1}{m} \prod_{i=2}^{k} \frac{(m-i-1)-1}{m-i-1}$$

The standard errors of reliability at time T are:

$$SE\left(\hat{R}_c(T)\right) \approx \hat{R}_c(T) \sqrt{\sum_{i=1}^{k} \frac{1}{(n-i)(n-i+1)}}$$

$$SE\left(\hat{R}_e(T)\right) \approx \hat{R}_c(T)\sqrt{\sum_{i=1}^{k}\frac{1}{(m-i)(m-i+1)}}$$

for k s.t. $t_k = T$ (If the life table method is used to calculate reliability estimates, use Greenwood's (1926) approximation formula for the standard error.)

Critical value(s):
    Reject H_0 if:

$$\hat{R}_e(T) - (1-\delta)\hat{R}_c(T) + t_\beta\sqrt{SE^2(\hat{R}_e(T)) + SE^2(\hat{R}_c(T))} \geq 0$$

where t_β = 100(1 − β) percentile of a (central) t distribution with $m + n - 4$ degrees of freedom.

Discussion:
    The rationale for this test is identical to that of Test 2.3, the one-sided alternate paradigm for comparing two independent means. Note that this test is referred to here as "nonparametric" because no particular parametric forms for the hazard functions or associated reliability functions are assumed.

Example:
    Suppose at $T = 0.5$,

$$\hat{R}_c(T) = 0.90$$

$$\hat{R}_e(T) = 0.87$$

$$m = n = 100$$

$$\delta = 0.025$$

$$SE\left(\hat{R}_c(T = 0.5)\right) \approx 0.03$$

$$SE\left(\hat{R}_e(T = 0.5)\right) \approx 0.03.$$

Then

$$\hat{R}_e(T) - (1-\delta)\hat{R}_c(T) + t_\beta\sqrt{SE^2(\hat{R}_e(T)) + SE^2(\hat{R}_c(T))}$$

$$= 0.87 - (1 - 0.025)0.90 + 1.653\sqrt{0.03^2 + 0.03^2} \approx 0.063 \geq 0.$$

Therefore, the null hypothesis of inferiority, H_0, is rejected in favor of noninferiority, H_1.

Confidence interval formulation:

$$\hat{R}_e(T) - (1-\delta)\hat{R}_c(T) + t_\beta \sqrt{SE^2(\hat{R}_e(T)) + SE^2(\hat{R}_c(T))}$$

is an approximate $100(1 - \beta)$ percent upper confidence limit on $R_e(T) - (1-\delta) R_c(T)$.

Computational considerations:
 This test is conceptually identical to Test 2.3.

Test 7.3 Accelerated Life Test with Type I Censoring

Parameters:
 λ_u = unaccelerated failure rate
 λ_a = accelerated failure rate
 f = acceleration parameter, that is, $\lambda_a = f\lambda_u$
 n = number of units tested
 T_c = censoring time, accelerated conditions
 $T_u = fT_c$ = time equivalent to T_c censoring time under unaccelerated conditions
 $\lambda_{u,0}$ = maximum tolerable failure rate, unaccelerated conditions
 $\lambda_{a,0}$ = maximum tolerable failure rate, accelerated conditions
 $\lambda_{a,0} = f\lambda_{u,0}$
 $r_0 = e^{-\lambda_{u,0}fT_c} = e^{-\lambda_{a,0}T_c}$ = reliability at time T_c (accelerated)

Hypotheses:

$$H_0: \lambda_u > \lambda_{u,0}$$

$$H_1: \lambda_u \leq \lambda_{u,0}$$

or equivalently

$$H_0: \lambda_a > f\lambda_{u,0}$$

$$H_1: \lambda_a \leq f\lambda_{u,0}$$

 Note that from the parameter list:

$$\lambda_{a,0} = \frac{-\ln r_0}{T_c}.$$

<u>Data:</u>

K = number of units failing at or before time T_c

x_i = time of ith failure, for the K units that failed at or before time T_c

$$\bar{T} = \sum_{i=1}^{K} \frac{x_i}{K} + \frac{(n-K)T_c}{K}.$$

If no units failed before time T_c, then $\bar{T} = T_c$.

$$\hat{\lambda}_a = \frac{1}{\bar{T}}$$

<u>Critical value(s):</u>

The same critical values used in Test 4.1 would apply; namely, if time to event, T, has an exponential distribution, then

$$\hat{\gamma} = \sum_{i=1}^{n} T_i = n\bar{T}$$

has a gamma distribution with parameters $\lambda_{a,0}$ and n. Often n is called the "shape" parameter, and $\lambda_{a,0}$ is called the "scale" parameter. Thus, if

$$G(\hat{\gamma} \mid n, \lambda_{a,0}) = \int_0^{\hat{\gamma}} \frac{\lambda_{a,0}^n}{\Gamma(n)} x^{n-1} e^{-\lambda_0 x} dx \geq \beta$$

then reject H_0. That is, $G(.)$ is the cumulative distribution function for a gamma-distributed random variable with parameters $\lambda_{a,0}$ and n. Similarly, find the value, call it γ_c, such that

$$G(\gamma_c \mid n, \lambda_{a,0}) = \int_0^{\gamma_c} \frac{\lambda_{a,0}^n}{\Gamma(n)} x^{n-1} e^{-\lambda_{a,0} x} dx = \beta$$

and reject H_0 if

$$\hat{\gamma} \geq \gamma_c.$$

One could also rely on the central limit theorem, and reject the null hypothesis if:

$$\bar{T} + t_{1-\beta} \frac{\bar{T}}{\sqrt{n}} \geq \frac{1}{\lambda_{a,0}}.$$

In this test, due to the assumption that some results may be censored, we are strictly relying on the assumption that time to failure is exponentially distributed, so that the maximum likelihood estimator of the standard deviation of time to failure is $\bar{T}$.

Discussion:

Inasmuch as this test is similar to Test 4.1, the power calculations are virtually identical. Of particular interest is the acceleration parameter, f. One method of deriving this value is based on the Arrhenious reaction rate law (Mann, Schafer, and Singpurwalla, 1974); namely:

$$f = \frac{\lambda_a}{\lambda_u} = \exp\left(-\frac{E}{K}\left(\frac{1}{\tau_a} - \frac{1}{\tau_u}\right)\right)$$

where τ_a is the temperature of acceleration, and τ_u is the "unaccelerated" temperature, or the nominal temperature to which the items under test are normally exposed, and at which the desired reliability at time T_u is r_0. Generally, the temperatures are expressed in degrees Kelvin. The constant $B = E/K$ is the energy of activation divided by Boltzmann's constant. It is a constant that is specific to the materials and associate chemical reactions that occur to attenuate the performance characteristics of the items under test. In some cases, the value of $B = E/K$ is known, usually from previously performed empirical investigation. In some cases, it may be desirable to perform some experiments to estimate the value of B. A simple and limited experiment would consist of an accelerated life test at two temperatures, say, τ_1 and τ_2, observing times to failure for n items under each temperature, computing the failure rate estimates for each sample:

$$\hat{\lambda}_k = \frac{1}{\bar{T}_k} \quad \text{for } k = 1, 2$$

and then computing the least squares estimate for B:

$$\hat{B} = -\frac{2\left[\dfrac{\ln\hat{\lambda}_1}{\tau_1} + \dfrac{\ln\hat{\lambda}_2}{\tau_2}\right] - \left[\dfrac{1}{\tau_1} + \dfrac{1}{\tau_2}\right]\left[\ln\hat{\lambda}_1 + \ln\hat{\lambda}_2\right]}{2\left[\left(\dfrac{1}{\tau_1}\right)^2 + \left(\dfrac{1}{\tau_2}\right)^2\right] - \left[\dfrac{1}{\tau_1} + \dfrac{1}{\tau_2}\right]^2}$$

which is based on the linear model for failure rate:

$$\ln(\lambda_{\omega i}) = \ln A - B\omega_i + v_i$$

where

$$\omega_i = \frac{1}{\tau_i}$$

and v_i represents model errors.

Example:

Suppose that under unaccelerated conditions, it is desired that at $T_u = 36$ hours, the reliability of a device is at least 9 percent, that is,

$$r_0 = e^{-\lambda_{u,0}T_u} = 0.90 \text{ so that}$$

$$\lambda_{u,0} = \frac{-\ln r_0}{T_u} = \frac{-\ln(0.90)}{36} \approx 0.002927$$

With an acceleration rate of 10,

$$\lambda_{a,0} = f\lambda_{u,0} = 10(0.002927) = 0.02927$$

$$\frac{1}{\lambda_{a,0}} = \frac{1}{0.02927} \approx 34.17$$

$$T_c = \frac{36}{10} = 3.6 \text{ hours.}$$

The hypotheses are

$$H_0: \lambda_a > f\lambda_{u,0} = 0.02927$$

$$H_1: \lambda_a \le f\lambda_{u,0} = 0.02927$$

Out of $n = 100$ items put on accelerated life test, 90 did not fail before 3.6 hours, and 10 failed at 1.5 hours, so that

$$\bar{T} = \sum_{i=1}^{K} \frac{x_i}{K} + \frac{(n-K)T_c}{K} = \frac{10(1.5)}{10} + \frac{(100-10)3.6}{10} \approx 33.9 \text{ hours.}$$

The test statistic is

$$\bar{T} + t_{1-\beta}\frac{\bar{T}}{\sqrt{n}} \approx 33.9 + 1.6604\left(\frac{33.9}{\sqrt{100}}\right) \approx 39.53 \ge \frac{1}{\lambda_{a,0}} \approx 34.17.$$

Therefore, the null hypothesis is rejected.

Confidence interval formulation:

$$\bar{T} + t_{1-\beta}\,\frac{\bar{T}}{\sqrt{n}}$$

is an approximate $100(1 - \beta)$ percent upper confidence limit on $1/\lambda_a$, so

$$f\left(\bar{T} + t_{1-\beta}\,\frac{\bar{T}}{\sqrt{n}}\right)$$

is an approximate $100(1 - \beta)$ percent upper confidence limit on $1/\lambda_u$.

Computational considerations:
    The code for this test is identical to the code for Test 4.1, except that the estimate of $\bar{T}$ accounts for censored data.

8

Sample-Based Criteria

In the previous seven chapters, critical values were always based on some hypotheses stated a priori and on desired risk levels for rejecting the null hypothesis. Unfortunately, sometimes a naïve approach is taken in choosing critical values for a test, without regard to any hypotheses, sample size, or associated risk levels. In this chapter, several such cases are discussed. Most of these cases are formulated as acceptance sampling problems as opposed to equivalence or noninferiority tests, per se.

Test 8.1 Single Proportion

<u>Data:</u>

$$X = \text{number of successes out of } n \text{ Bernoulli trials}$$

$$\hat{P} = \frac{X}{n}$$

<u>Critical value(s):</u>

Pass the test if $\hat{P} \geq p_c$.

<u>Discussion:</u>

Given that the population proportion of successes, P, is unknown, and there is no hypothesis concerning its minimum acceptable value or maximum unacceptable value, it is not possible to specify the risk of failing (or passing). As a consequence, it is not possible to determine what sample size, n, would be required. If the sample size is fixed to some arbitrary value, then a risk curve for that sample size could be constructed. In other words, the critical number of successes would be:

$$X_c = round\,(np_c,\, 0)$$

where $round(x, 0)$ means round x to the nearest integer.

For $P = 0$ to 1, the probability of passing the test is

$$\Pr\{X \geq X_c \mid P, n\} = \sum_{k=X_c}^{n} \binom{n}{k} (P)^k (1-P)^{n-k}.$$

TABLE 8.1

Probabilities of Passing the Sample-
Based Criterion Test for Proportions

N	P	X_c = 95% of N	$Pr\{X \geq X_c\}$
200	0.95	190	58.31%
100	0.95	95	61.60%
50	0.95	48	54.05%
200	0.97	190	95.99%
100	0.97	95	91.92%
50	0.97	48	81.08%

However, since no particular minimum acceptable probability of success was specified, it is not possible to determine whether the sample size, n, is sufficient, too small, or even too large.

Example:

Suppose $n = 100$, and the criterion for passing the test is that $\hat{P} \geq 0.95$, so that $X_c = round\,(100*0.95, 0) = 95$. If the population's probability of a "success" is $P = 0.95$, then the probability of passing the criterion is:

$$\Pr\{X \geq 95\,|\,0.95, 100\} = \sum_{k=95}^{100} \binom{100}{k}(0.95)^k\,(0.05)^{100-k} \approx 0.616$$

or about a 61.6 percent chance. Conversely, suppose that the sample size is changed to 50, but the sample criterion remains the same, that $\hat{P} \geq 0.95$ in order to pass. Then, under the same population condition, that is, $P = 0.95$, the probability of passing is now:

$$\Pr\{X \geq 48\,|\,0.95, 48\} = \sum_{k=48}^{50} \binom{50}{k}(0.95)^k\,(0.05)^{50-k} \approx 0.540$$

or about a 54.0 percent chance of passing, even though the sample criterion is still 95 percent of n. Table 8.1 shows the probabilities of "passing" the test for $n = 200$, 100, and 50, under the conditions that $P = 0.95$ and $P = 0.97$.

Test 8.2 Single Mean

Data:

$$\bar{X} = \text{sample mean}$$

$$n = \text{sample size}$$

Critical value(s):
  Pass the test if:

$$\bar{X} \le \bar{x}_c.$$

Discussion:
  The central limit theorem dictates that the sample statistic, $\bar{X}$, tends toward having a normal distribution, that is,

$$\bar{X} \sim N\left(\mu, \frac{\sigma^2}{n}\right).$$

Clearly, the probability of passing the test depends on the parameters μ, σ, and n. Without specifying μ and σ, it is not possible to assess the risk curve for such a test. Of course, as in the case of the single proportion, fixing the acceptance criterion/critical value regardless of sample size results in risks that potentially change for each instance of the test, if the sample size is not fixed.

Example:
  Suppose that the pass criterion is that:

$$\bar{X} \le \bar{x}_c = 100.$$

Suppose further that $\mu = 95$, $\sigma = 14$, and $n = 30$. Then

$$\Pr\left\{\bar{X} \le 100 \mid \mu = 95, \sigma = 14, n = 30\right\} = \frac{\sqrt{30}}{14\sqrt{2\pi}} \int_{-\infty}^{100} e^{-\frac{1}{2}\left(\frac{x-95}{14/\sqrt{30}}\right)^2} dx \approx 0.9748.$$

If the sample size were changed to $n = 10$, without changing the critical value, the associated probability of passing would be:

$$\Pr\left\{\bar{X} \le 100 \mid \mu = 95, \sigma = 14, n = 10\right\} = \frac{\sqrt{10}}{14\sqrt{2\pi}} \int_{-\infty}^{100} e^{-\frac{1}{2}\left(\frac{x-95}{14/\sqrt{10}}\right)^2} dx \approx 0.8706.$$

Test 8.3 Relative Difference between Two Means

Data:

$$\bar{X}_a = \text{sample mean of antecedent system}$$

$$\bar{X}_d = \text{sample mean of descendent system}$$

Critical value(s):
    Pass the test if:

$$\frac{\bar{X}_d - \bar{X}_a}{\bar{X}_a} \leq r.$$

In other words, it is desirable for the descendent system mean to be no more than 100r percent greater than the mean of the antecedent system.

Discussion:
    Even under the restriction that the sample means are independent, the sampling distribution of the ratio:

$$\frac{\bar{X}_d - \bar{X}_a}{\bar{X}_a}$$

is at best complicated, and beyond the ability of most statistical analysis programs to compute. Consequently, the risk characteristics of this test could only be computed using a specially designed and written computer program (Springer, 1979). Alternatively, a Monte Carlo simulation approach could be used if the parameters μ_a, σ_a, μ_d, σ_d, and n were specified. However, it must be emphasized that:

$$E\left[\frac{\bar{X}_d - \bar{X}_a}{\bar{X}_a}\right] \neq \frac{\mu_d - \mu_a}{\mu_a}$$

and in fact this expectation is not finite.

Example:
    The "pass" criterion is that:

$$\frac{\bar{X}_d - \bar{X}_a}{\bar{X}_a} \leq 0.03.$$

If $n = 30$ for both antecedent and descendent systems, then there is an infinite number of possible parameter vectors $[\mu_a, \sigma_a, \mu_d, \sigma_d]$ for which

$$\sup \Pr\left\{\frac{\bar{X}_d - \bar{X}_a}{\bar{X}_a} \leq 0.03\right\} = 1 - \beta = 0.95$$

even with the sample sizes fixed. Furthermore, the condition:

$$\frac{\mu_d - \mu_a}{\mu_a} = 0.03$$

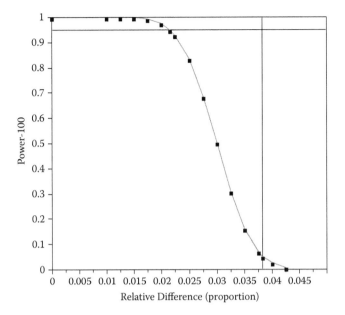

FIGURE 8.1
Test 8.3, power curve for relative difference of means, $n = 30$ per system, $\mu_a = 100$.

is neither necessary nor sufficient to ensure any particular value of $1 - \beta$, including our customary $1 - \beta = 0.95$. A simulation was run in which $\sigma_a = \sigma_d = 2.0$, $\mu_a = 100$, and $n = 30$ for each system. Figure 8.1 shows the power curve for the test.

Thus, in this case, if we had decided to make the hypothesis test for

$$H_0: \frac{\mu_d - \mu_a}{\mu_a} > 0.0216$$

$$H_1: \frac{\mu_d - \mu_a}{\mu_a} \le 0.0216$$

then under the conditions stated above for the parameters, with the sample criterion that we would reject H_0 if

$$\frac{\bar{X}_d - \bar{X}_a}{\bar{X}_a} \le 0.03$$

there would be approximately a $1 - \beta = 0.95$ chance of rejecting H_0 if it were true that $\frac{\mu_d - \mu_a}{\mu_a} = 0.0215$.

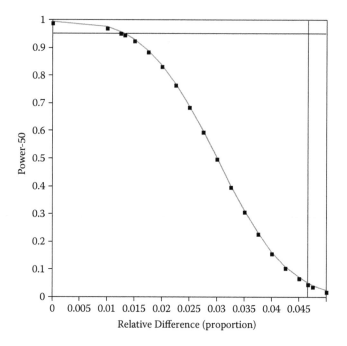

FIGURE 8.2
Test 8.3, power curve for relative difference of means, $n = 30$ per system, $\mu_a = 50$.

If instead of $\mu_a = 100$, we had $\mu_a = 50$, the risk curve would have changed. Figure 8.2 shows the curve for this case.

Now, with $\mu_a = 50$, the hypotheses having the same power to reject are changed to:

$$H_0: \frac{\mu_d - \mu_a}{\mu_a} > 0.0132$$

$$H_1: \frac{\mu_d - \mu_a}{\mu_a} \leq 0.0132.$$

Even though the sample sizes are identical, the assumptions about the standard deviations are identical, and the critical values for the test statistic are identical, the hypotheses that generate the same level of risk (power) are very different. With a critical value fixed without regard to sample size or population parameters, it is not possible to specify a "null" value for the parameter

$$\frac{\mu_d - \mu_a}{\mu_a}$$

without also specifying a value for μ_a. By fixing the critical value of the test statistic, we can no longer specify how close the descendent population mean must be to the antecedent in order to be close enough. Under the paradigm that $\mu_a = 100$, in this case, we have implied that 2.16 percent is close enough. Under the paradigm that $\mu_a = 50$, we have implied that 1.32 percent is close enough. The situation begs the questions:

1. If we care about relative percent difference, then why would we not have a single threshold for μ_d in terms of percent difference from μ_a?
2. If we really care about relative percent difference only, why do we not test the hypotheses as described in Chapter 2, Test 2.3, so that the population relative percent error is constant, regardless of the values of the antecedent mean?

Test 8.4 Single Standard Deviation

Data:

$$S = \text{sample standard deviation} = \sqrt{\frac{\sum_{i=1}^{n}(x_i - \bar{x})^2}{n-1}}$$

$$n = \text{sample size}$$

Critical value(s):
Pass the test if $S \leq s_c$.

Discussion:
From Test 3.1, "passing" implies that

$$S^2 \leq s_c^2 = \frac{\sigma_0^2}{n-1}\chi_{1-\beta}^2(n-1)$$

where $\chi_{1-\beta}^2(n-1) = 100(1-\beta)$ percentile of a chi-squared distribution with $n-1$ degrees of freedom, but σ_0, the population standard deviation, is unspecified. Thus, in order for this test to have a $100(1-\beta)$ percent chance of resulting in a "pass,"

$$\sigma_0 = \sqrt{\frac{(n-1)s_c^2}{\chi_{1-\beta}^2(n-1)}}.$$

Clearly, if the sample size changes but s_c remains constant, then the value of σ_0 required to yield a $100(1 - \beta)$ percent chance of "passing" changes. Thus, by fixing the critical value, s_c, without adjusting for sample size, the risk characteristics of the test change, potentially in an undesirable fashion.

Example:
Suppose we decide to sample $n = 30$ items, and reject if $S \leq s_c = 1.0$. In order to have a $100(1 - \beta)$ percent $= 95$ percent chance of "passing," we would require

$$\sigma_0 = \sqrt{\frac{(30-1)(1.0)^2}{\chi^2_{.05}(30-1)}} \approx 0.8255.$$

Conversely, if $n = 15$ with the same criterion, $S \leq s_c = 1.0$, the implication is that

$$\sigma_0 = \sqrt{\frac{(15-1)(1.0)^2}{\chi^2_{.05}(15-1)}} \approx 0.7688.$$

Figure 8.3 shows a plot of values of σ_0 as a function of sample size, n, with $s_c = 1.0$. Thus, with $n = 30$, the fixed sample criterion implies that we are willing to tolerate a population standard deviation as high as $\sigma_0 = 0.8255$, but with $n = 15$, we are only willing to tolerate a population standard deviation as high as $\sigma_0 = 0.7688$.

In general, specifying a constant sample acceptance criterion, regardless of sample size or population criteria, is like marking the finish line for a race, but changing the starting line every time the race is run.

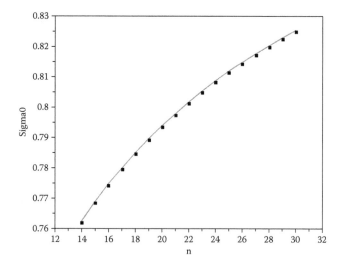

FIGURE 8.3
Test 8.4, maximum tolerable sigma implied by $S \leq s_c$, for s_c fixed regardless of n.

9

Comparing Sequences of Points

Test 9.1 Two Sequences, Variability Known

Parameters:

$$f_e(t) = \mu_e(t) + \varepsilon, \quad \varepsilon \sim N(0, \sigma_e^2) \; \forall t$$

$$f_r(t) = \mu_r(t) + \eta, \quad \eta \sim N(0, \sigma_r^2) \; \forall t$$

Note that $\mu_e(t)$ and $\mu_r(t)$ are deterministic functions of t.
σ_e^2 = variance of noise in $f_e(t)$ (assumed known)
σ_r^2 = variance of noise in $f_r(t)$ (assumed known)
δ_0 = maximum desirable difference between the sequences

Hypotheses:

$$H_0: |\mu_e(t) - \mu_r(t)| > \delta_0 \; \forall t$$

$$H_1: |\mu_e(t) - \mu_r(t)| \le \delta_0 \; \forall t$$

Data:

t_i = the ith value of a sequencing variable (often time), for $i = 1, n$

$f_e(t_i)$ = observation of one of two sequences sampled at t_i

$f_r(t_i)$ = observation of the other of two sequences sampled at t_i

$$\psi^2 = \sum_{i=1}^{n} \left(f_e(t_i) - f_r(t_i) \right)^2 = \text{the sum of squared differences between the}$$

two sequences

Critical value(s):
  Reject H_0 if

$$\frac{\psi^2}{\sigma_e^2 + \sigma_r^2} \le \chi_{1-\beta}^{\prime 2}\left(n, \lambda_0 = \frac{n\delta_0^2}{\sigma_e^2 + \sigma_r^2} \right)$$

$$\chi_{1-\beta}^{\prime 2}\left(n, \lambda_0 = \frac{n\delta_0^2}{\sigma_e^2 + \sigma_r^2}\right)$$

is the $100(1 - \beta)$ percentile of a noncentral chi-squared distribution with n degrees freedom and noncentrality parameter λ_0.

Discussion:
  Note that:

$$\psi^2 = \sum_{i=1}^{n}\left(f_e(t_i) - f_r(t_i)\right)^2$$

$$= \left(\sigma_e^2 + \sigma_r^2\right)\sum_{i=1}^{n}\frac{\left(f_e(t_i) - f_r(t_i) - \left(\mu_e(t_i) - \mu_r(t_i)\right) + \left(\mu_e(t_i) - \mu_r(t_i)\right)\right)^2}{\left(\sigma_e^2 + \sigma_r^2\right)}.$$

If:

$$\mu_e(t_i) - \mu_r(t_i) = \delta_0 \quad \forall i$$

then

$$\frac{\psi^2}{\sigma_e^2 + \sigma_r^2} \sim \chi^{\prime 2}\left(n, \lambda_0 = \frac{n\delta_0^2}{\sigma_e^2 + \sigma_r^2}\right)$$

a noncentral chi-squared with noncentrality

$$\lambda_0 = \frac{n\delta_0^2}{\sigma_e^2 + \sigma_r^2}$$

(Johnson, Kotz, and Balakrishnan, 1995). As the difference $\mu_e(t_i) - \mu_r(t_i)$ gets larger than δ_0, the chance that the statistic

$$\frac{\psi^2}{\sigma_e^2 + \sigma_r^2}$$

is smaller than the critical value

$$\chi_{1-\beta}^{\prime 2}\left(n, \lambda_0 = \frac{n\delta_0^2}{\sigma_e^2 + \sigma_r^2}\right)$$

diminishes.

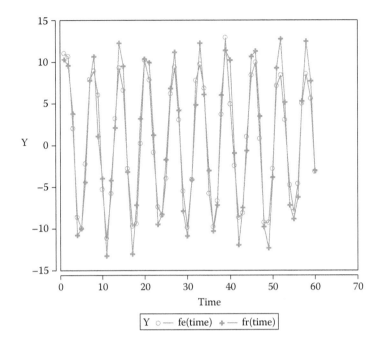

FIGURE 9.1
Two sequences of points.

Example:
  Consider the two sequences of values shown in Figure 9.1.
  Figure 9.2 shows the differences, point by point.
  Assume that $\sigma_e = \sigma_r = 1.665$, so that $\sigma_e^2 + \sigma_r^2 = 3.33$. From these data ($n = 60$ per sequence):

$$\psi^2 = \sum_{i=1}^{n} \left(f_e(t_i) - f_r(t_i) \right)^2 \approx 280.6508.$$

Thus,

$$\frac{\psi^2}{\sigma_e^2 + \sigma_r^2} \approx 84.2795.$$

If we were testing the hypothesis:

$$H_0: |\mu_e(t) - \mu_r(t)| > \delta_0 = 0.33 \quad \forall t$$

against the hypothesis:

$$H_1: |\mu_e(t) - \mu_r(t)| \leq 0.33 \quad \forall t$$

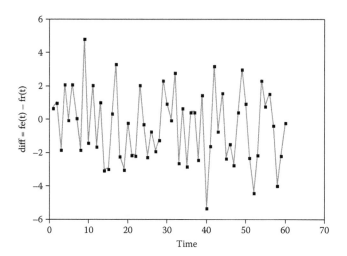

FIGURE 9.2
Differences, $f_e(t_i) - f_r(t_i)$.

the critical value, with $1 - \beta = 0.95$, is

$$\chi^{'2}_{0.95}\left(n = 60, \lambda_0 = \frac{60(0.33)^2}{3.33}\right) \approx 81.6569 .$$

Since the test statistic is greater than the critical value (84.2795 > 81.6569), we fail to reject the null hypothesis and conclude that the sequences are not equivalent. Figure 9.3 shows the power curve for this test.

Confidence interval formulation:
    An upper $100(1 - \beta)$ percent confidence limit on

$$\delta = sup|\mu_e(t) - \mu_r(t)|$$

can be derived as follows. First, find the value, λ_u, the largest value of the noncentrality parameter such that

$$F\left(\frac{\psi^2}{\sigma_e^2 + \sigma_r^2}\,|\,n, \lambda_u\right) = 1 - \beta$$

where $F(.)$ is the cumulative distribution function of the noncentral chi-squared with n degrees of freedom. Then compute:

$$\delta_u = \left[\frac{(\sigma_e^2 + \sigma_r^2)\lambda_u}{n}\right]^{1/2}$$

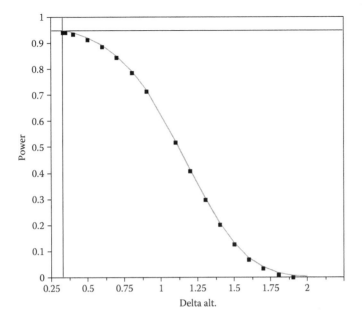

FIGURE 9.3
Test 9.1, power curve.

which is an upper $100(1 - \beta)$ percent confidence limit for δ.

Computational considerations:

Inasmuch as the next test, Test 9.2, is more generally applicable, and the methods are very similar, no code or computer output will be demonstrated for Test 9.1. However, computing applications are given for Test 9.2.

Test 9.2 Two Sequences, Variability Unknown

Parameters:

$$f_e(t) = \mu_e(t) + \varepsilon, \varepsilon \sim N\left(0, \sigma_e^2\right) \quad \forall t$$

$$f_r(t) = \mu_r(t) + \eta, \eta \sim N\left(0, \sigma_r^2\right) \quad \forall t$$

Note that $\mu_e(t)$ and $\mu_r(t)$ are deterministic functions of t.

σ_e^2 = variance of noise in $f_e(t)$ (assumed unknown)

σ_r^2 = variance of noise in $f_r(t)$ (assumed unknown)

δ_0 = maximum desirable difference between the sequences

$$\gamma_0 = \frac{\delta_0}{\sqrt{\sigma_e^2 + \sigma_r^2}} = \text{maximum desirable difference between the sequences,}$$

in standard deviation units

Hypotheses:

$$H_0 : \frac{|\mu(t) - \mu(t)|}{\sqrt{\sigma_e^2 + \sigma_r^2}} > \gamma_0 = \frac{\delta_0}{\sqrt{\sigma_e^2 + \sigma_r^2}} \quad \forall t$$

$$H_1 : \frac{|\mu(t) - \mu(t)|}{\sqrt{\sigma_e^2 + \sigma_r^2}} \leq \gamma_0 = \frac{\delta_0}{\sqrt{\sigma_e^2 + \sigma_r^2}} \quad \forall t$$

Data:

t_i = the ith value of a sequencing variable (often time), for $i = 1, n$

$f_e(t_i)$ = observation of one of two sequences sampled at t_i

$f_r(t_i)$ = observation of the other of two sequences sampled at t_i

$$\psi^2 = \sum_{i=1}^{n} \left(f_e(t_i) - f_r(t_i) \right)^2 = \text{the sum of squared differences between the}$$

two sequences

$\hat{\sigma}_e^2 + \hat{\sigma}_r^2$ = an estimate of the variance of paired differences between the two sequences

Critical value(s):
    Reject H_0 if:

$$\frac{\psi^2}{\hat{\sigma}_e^2 + \hat{\sigma}_r^2} \leq F_{1-\beta}' \left(n, n, \lambda_0 = \frac{n\delta_0^2}{\sigma_e^2 + \sigma_r^2} \right)$$

where

$$F_{1-\beta}' \left(n, n, \lambda_0 = \frac{n\delta_0^2}{\sigma_e^2 + \sigma_r^2} \right)$$

is the $100(1-\beta)$ percent percentile of a noncentral F-distribution with n degrees of freedom in the numerator and denominator, and numerator noncentrality

$$\lambda_0 = \frac{n\delta_0^2}{\sigma_e^2 + \sigma_r^2} = n\gamma_0^2.$$

In other words, since variability is unknown, the noncentrality parameter could be specified in standard deviation (of the difference) units.

Discussion:
  If

$$Var\left(f_e\left(t_i\right)-f_r\left(t_i\right)\right)=Var\left(D_i\right)=Var\left(\varepsilon-\eta\right)=\sigma_e^2+\sigma_r^2$$

then the maximum likelihood estimate of the variance of the difference $f_e(t_i)-f_r(t_i)$ is

$$\hat{\sigma}_e^2+\hat{\sigma}_r^2=\frac{1}{n}\sum_{i=1}^{n}\left(D_i-\bar{D}\right)^2$$

and

$$\bar{D}=\frac{1}{n}\sum_{i=1}^{n}D_i.$$

The test statistic

$$\frac{\psi^2}{\hat{\sigma}_e^2+\hat{\sigma}_r^2}$$

has a noncentral F-distribution with n degrees of freedom in both numerator and denominator, and noncentraility:

$$\lambda_0=\frac{n\delta_0^2}{\sigma_e^2+\sigma_r^2}=n\gamma_0^2$$

(Johnson, Kotz, and Balakrishnan, 1995).

Example:
  From the example for Test 9.1,

$$\psi^2=\sum_{i=1}^{n}\left(f_e\left(t_i\right)-f_r\left(t_i\right)\right)^2\approx 280.6508.$$

The estimate of noise variance is

$$\hat{\sigma}_e^2+\hat{\sigma}_r^2=\frac{1}{n}\sum_{i=1}^{n}\left(D_i-\bar{D}\right)^2\approx 4.96.$$

The test statistic is

$$\frac{\psi^2}{\hat{\sigma}_e^2+\hat{\sigma}_r^2}\approx\frac{280.6508}{4.9582}\approx 56.6034.$$

With:

$$\gamma_0 = \frac{\delta_0}{\sqrt{\sigma_e^2 + \sigma_r^2}} = \frac{0.33}{\sqrt{3.33}} \approx 0.1808$$

and $n = 60$, the "null" noncentrality is

$$\lambda_0 = \frac{n\delta_0^2}{\sigma_e^2 + \sigma_r^2} = n\gamma_0^2 \approx 60\left(.1808^2\right) \approx 1.9622.$$

The critical value from the noncentral F-distribution with $n = 60$ degrees of freedom in both numerator and denominator is approximately 1.5843. Since 56.6034 > 1.5843, the null is not rejected. The power curve for this example is given in Figure 9.4.

<u>Confidence interval formulation:</u>
An upper $100(1 - \beta)$ percent confidence limit on

$$\gamma = sup \frac{\left|\mu_e(t) - \mu_r(t)\right|}{\sqrt{\sigma_e^2 + \sigma_r^2}}$$

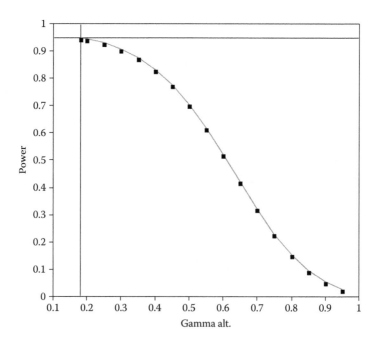

FIGURE 9.4
Test 9.2, power curve for comparing two sequences, variability unknown.

can be derived as follows. First, find the value, λ_u, the largest value of the noncentrality parameter such that

$$F\left(\frac{\psi^2}{\hat{\sigma}_e^2 + \hat{\sigma}_r^2} \mid n, n, \lambda_u\right) = 1 - \beta$$

where $F(.)$ is the cumulative distribution function of the noncentral F with n degrees of freedom in both numerator and denominator. Then compute:

$$\gamma_u = \left[\frac{\lambda_u}{n}\right]^{1/2}$$

which is an upper $100(1 - \beta)$ percent confidence limit for γ.

Computational considerations:

- SAS code

```
libname stuff 'H:\Personal Data\Equivalence & Noninferiority\
Programs & Output';

data calc;

  set stuff.d20121116_test_9_2_example_data;
  diff = fe - fc;
  diffsq = diff**2;
  run;

proc print data = calc;
  run;

proc means data = calc;
  var diff diffsq;
  output out = onemean SUM = sdiff sdiffsq VAR = vdiff N =
  ndiff;
  run;

data outcalc;

  set onemean;
  delsig = 5;
  nc0 = ndiff*delsig**2;
  beta = 0.05;
  psisq = sdiffsq/vdiff;
  f_crit = finv(1-beta,ndiff,ndiff,nc0);
  run;
```

```
proc print data = outcalc;/* has vars sdiff sdiffsq vdiff
ndiff delsig nc0 beta psisq f_crit */

   run;
```

The SAS System   11:02 Friday, November 16, 2012 3

Obs	time	fc	fe	diff	diffsq
1	1	10.0000	11.0749	1.07488	1.15536
2	2	5.9697	5.0135	-0.95619	0.91431
3	3	1.5610	1.6425	0.08156	0.00665
4	4	5.7207	5.8904	0.16973	0.02881
5	5	8.5700	8.4026	-0.16740	0.02802
6	6	9.8114	9.3124	-0.49895	0.24895
7	7	13.4625	13.9105	0.44803	0.20073
8	8	17.4663	16.6408	-0.82550	0.68145
9	9	19.5568	20.0469	0.49017	0.24027
10	10	18.9278	20.6939	1.76616	3.11933
11	11	22.5127	23.0164	0.50366	0.25367
12	12	22.4253	23.7516	1.32634	1.75918
13	13	14.7594	13.0602	-1.69920	2.88727
14	14	10.6493	11.0431	0.39383	0.15510
15	15	11.3332	10.6330	-0.70021	0.49029
16	16	8.2396	7.9960	-0.24355	0.05932
17	17	9.5967	8.8327	-0.76401	0.58371
18	18	10.6682	12.3575	1.68934	2.85388
19	19	12.2426	12.3369	0.09434	0.00890
20	20	17.9369	19.3441	1.40723	1.98031
21	21	12.4793	12.8372	0.35789	0.12808
22	22	11.5343	12.3369	0.80261	0.64418
23	23	16.0164	14.3267	-1.68972	2.85514
24	24	18.7508	18.9056	0.15478	0.02396
25	25	19.8442	19.3896	-0.45452	0.20659
26	26	18.9532	19.5501	0.59686	0.35624
27	27	12.0322	12.8241	0.79197	0.62721
28	28	16.4105	17.6849	1.27436	1.62401
29	29	8.5825	7.1428	-1.43971	2.07277
30	30	2.1924	2.2532	0.06076	0.00369
31	31	-1.5464	-3.1166	-1.57019	2.46550
32	32	-4.9271	-4.3207	0.60640	0.36772
33	33	-4.8922	-5.6441	-0.75185	0.56528
34	34	-7.9586	-9.0416	-1.08294	1.17276
35	35	-8.2935	-7.9490	0.34451	0.11868
36	36	-1.0380	-0.6618	0.37619	0.14152
37	37	-3.5019	-1.9691	1.53274	2.34928
38	38	-9.2971	-8.9091	0.38805	0.15058
39	39	-8.2664	-9.5914	-1.32503	1.75570
40	40	-10.0305	-9.7129	0.31761	0.10088

```
        The SAS System          11:02 Friday, November 16, 2012 4

                          The MEANS Procedure

Variable     N        Mean       Std Dev      Minimum      Maximum
ƒƒƒƒƒƒƒƒƒƒƒƒƒƒƒƒƒƒƒƒƒƒƒƒƒƒƒƒƒƒƒƒƒƒƒƒƒƒƒƒƒƒƒƒƒƒƒƒƒƒƒƒƒƒƒƒƒƒƒƒƒƒƒƒ
diff         40    0.0720266    0.9497339   -1.6991980    1.7661628
diffsq       40    0.8846325    0.9877772    0.0036921    3.1193311
ƒƒƒƒƒƒƒƒƒƒƒƒƒƒƒƒƒƒƒƒƒƒƒƒƒƒƒƒƒƒƒƒƒƒƒƒƒƒƒƒƒƒƒƒƒƒƒƒƒƒƒƒƒƒƒƒƒƒƒƒƒƒƒƒ

        The SAS System          11:02 Friday, November 16, 2012 5

Obs _TYPE_ _FREQ_  sdiff  sdiffsq  vdiff  ndiff delsig nc0  beta  psisq   f_crit
 1    0      40   2.88106 35.3853 0.90199  40     5   1000 0.05 39.2301 39.6493
```

Since psisq < f_crit, the null hypothesis is rejected in favor of the alternative; namely, the two sequences are equivalent.

- JMP Data Table and formulas (Figures 9.5, 9.6, and 9.7)

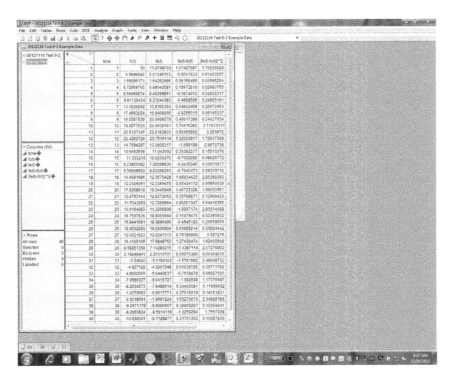

FIGURE 9.5
Test 9.2, JMP screen 1.

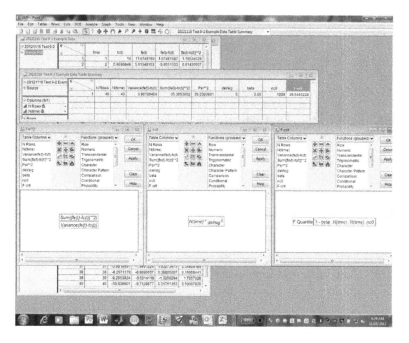

FIGURE 9.6
Test 9.2, JMP screen 2.

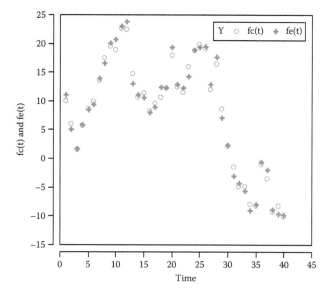

FIGURE 9.7
Plot of the two sequences.

10

Calculating Sample Sizes

<u>Parameters and variables:</u>
- n = sample size
- θ = population parameter about which the inference is to be made
- T = test statistic, which is a function of the observed data, and whose probability distribution depends on the value of θ and n
- T_c = critical value of the test statistic, being such that if the test statistic exceeds this value, the null is rejected
- θ_0 = the value of the parameter that defines the boundary between acceptable and unacceptable conditions
- ψ = prespecified parameters other than the one(s) for which an inference (hypothesis test) is to be made. For example, in a test about means, standard deviation, σ, is prespecified for purposes of power calculations.

<u>Discussion:</u>
In general, the probability of rejecting the null hypothesis can be expressed as:

$$\Pr\{|T| \geq T_c|\theta, n, \psi\}.$$

We generally have chosen the critical value, T_c, so that:

$$\sup \Pr\{|T| \geq T_c|\theta_0, n, \psi\} = 1 - \beta.$$

Suppose we could choose another potential value of the parameter θ, say, θ_a, such that:

$$\inf \Pr\{T \geq T_c|\theta_a, n, \psi\} = \alpha$$

for some specified value of $\alpha < 1 - \beta$. Then, in theory, the two equations could be used to solve for the sample size, n. There are some pragmatic issues associated with this methodology. In particular, it is often difficult for experimenters to specify the value of θ_a. It may, however, be easier for experimenters to determine a potential range of economically feasible sample sizes (or at least an upper bound). Thus, rather than calculating the solution to the two simultaneous equations, the value of n might be fixed, the critical value T_c determined using the first equation, and then the value of θ_a determined for a fixed value of α. The value of θ_a determined in this fashion may

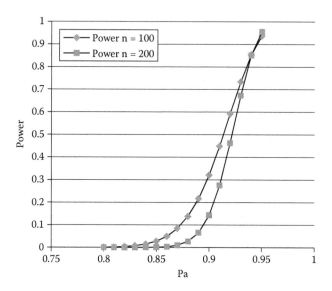

FIGURE 10.1
Power curves for tests of a single proportion (Test 1.1).

shock the experimenter into choosing a larger sample size, or it may seem adequate. The question of "How bad is too bad?" is very context sensitive, as is the question of "How large a sample size can we afford?"

For the tests described in this work, as sample size increases, the power to reject the null hypothesis actually decreases. While this may seem counterintuitive, it is in fact the appropriate relationship between power and sample size. Larger sample sizes will result in more stringent tests. This is true in regard to conventional hypothesis tests. Figure 10.1 shows the power curves for a test of a single proportion (Test 1.1), with sample sizes $n = 100$ and $n = 200$. The hypotheses are:

$$H_0: P < 0.95$$

$$H_1: P \geq 0.95.$$

Note, for example, if $P = 0.90$, with $n = 100$, the power to reject the null is approximately 0.3209 (32.09 percent). With $n = 200$, the power at $P = 0.90$ is 0.1431 (14.31 percent). Thus, when the sample size doubled, the power was reduced to less than half. Similarly, Figure 10.2 shows power curves for a test of a single mean (Test 2.1) with $n = 20$ and $n = 40$. The hypotheses are

$$H_0: \mu < 100$$

$$H_1: \mu \geq 100$$

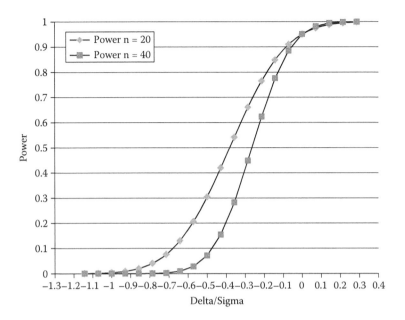

FIGURE 10.2
Power curves for tests of a single mean (Test 2.1).

The graph is in terms of:

$$\frac{\delta}{\sqrt{n}} = \frac{(\mu_a - 100)}{\sigma}$$

or the difference between the (hypothetical alternative) population mean and the null value of 100, in standard deviation units. At

$$\frac{(\mu_a - 100)}{\sigma} = -0.5$$

with $n = 20$, there is approximately a power to reject the null of 0.3048 (30.48 percent). With $n = 40$, the power is approximately 0.0719 (7.19 percent).

The reader is referred to Desu and Raghavarao (1990) for a more general discussion of sample size calculation methods.

11

Computer Code—Power Curves

This chapter describes computer codes and procedures for generating the power curves for various tests. Each test (except Test 6.1, for which only R code is presented) will have the following parts:

Input parameters
SAS code
JMP Data Table and formulas
R

Test 1.1 Single Binomial Proportion (One-Sided—Probability of "Success")

Input parameters:
P0 = minimum desired probability of success
P1 = alternate values for probability of success (P1 ≤ P0)
N = sample size
Xc = critical value for number of successes (yields ~ 1 − beta probability of rejecting the null hypothesis if probability of success = P_0)

SAS code:

```
libname stuff 'H:\Personal Data\Equivalence & Noninferiority\
Programs & Output';

data calc;
  set stuff.D20120730_One_Proportion_Power;
  power = 1 - probbnml(P1,N,Xc-1);/* Pr{X > = Xc} */
  run;

proc print data = calc;/*dataset calc has columns N, Xc, P0,
P1, power */

  run;
```

Output:

```
        The SAS System        09:04 Monday, July 30, 2012 1

                    Obs     N     Xc      P0      P1      power
                     1     100    92    0.95    0.95    0.93691
                     2     100    92    0.95    0.94    0.85371
                     3     100    92    0.95    0.93    0.73397
                     4     100    92    0.95    0.92    0.59263
                     5     100    92    0.95    0.91    0.44940
                     6     100    92    0.95    0.90    0.32087
                     7     100    92    0.95    0.89    0.21651
                     8     100    92    0.95    0.88    0.13859
                     9     100    92    0.95    0.87    0.08448
                    10     100    92    0.95    0.86    0.04921
                    11     100    92    0.95    0.85    0.02748
                    12     100    92    0.95    0.84    0.01474
                    13     100    92    0.95    0.83    0.00762
                    14     100    92    0.95    0.82    0.00380
                    15     100    92    0.95    0.81    0.00183
                    16     100    92    0.95    0.80    0.00086
```

JMP Data Table and formulas:

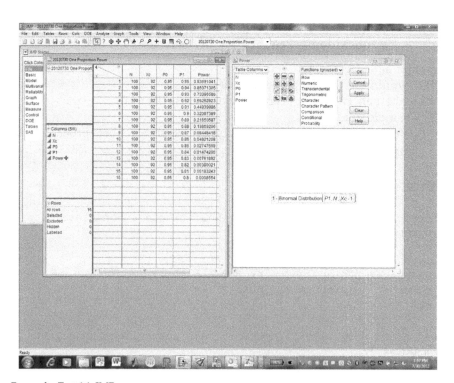

Power for Test 1.1, JMP screen.

Note that the column labeled "Power" is computed as described in the formula view.

R:

```
> prob <- c(0.95,.94,.93,.92,.91,.90,.89,.88,.87,.86,.85,.84,
.83,.82,.81,.80)
> power <- 1 - pbinom(92,100,prob)
> power
[1]  0.8720395214  0.7483494398  0.5987792361  0.4471097512
0.3127949596
[6]  0.2060508618  0.1284536668  0.0761360984  0.0430807590
0.0233537817
[11]  0.0121651919  0.0061048616  0.0029577572  0.0013860090
0.0006291400
[16]  0.0002769869
> plot(prob,power)
```

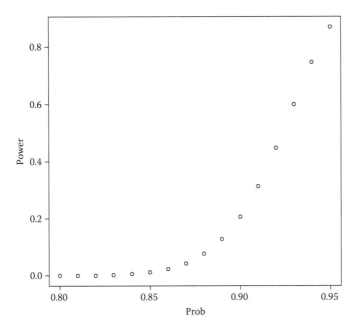

Test 1.1, R: power curve.

Test 2.1 Single Mean (One-Sided)

<u>Input parameters:</u>
 muA = alternative values of the mean
 muL = minimum desirable value for the mean

sigma = prior estimate of the standard deviation
n = sample size

SAS code:

```
libname stuff 'H:\Personal Data\Equivalence & Noninferiority\
Programs & Output';

data calc;
  set stuff.d20120801_one_mean;
  delta = (muA - muL)/sigma;
  nc = sqrt(n) * delta;
  tcrit = tinv(0.05,n-1);
  power = 1 - probt(tcrit,n-1,nc);/* Pr{T' > = tcrit} */
  run;

proc print data = calc;/*dataset calc has columns n muL muA
sigma delta nc power */

  run;
```

The SAS System   09:11 Wednesday, August 1, 2012 1

Obs	n	muL	muA	sigma	delta	nc	tcrit	power
1	20	100	96.00	3.5	-1.14286	-5.11101	-1.72913	0.00053
2	20	100	96.25	3.5	-1.07143	-4.79157	-1.72913	0.00150
3	20	100	96.50	3.5	-1.00000	-4.47214	-1.72913	0.00390
4	20	100	96.75	3.5	-0.92857	-4.15270	-1.72913	0.00929
5	20	100	97.00	3.5	-0.85714	-3.83326	-1.72913	0.02033
6	20	100	97.25	3.5	-0.78571	-3.51382	-1.72913	0.04094
7	20	100	97.50	3.5	-0.71429	-3.19438	-1.72913	0.07597
8	20	100	97.75	3.5	-0.64286	-2.87494	-1.72913	0.13021
9	20	100	98.00	3.5	-0.57143	-2.55551	-1.72913	0.20668
10	20	100	98.25	3.5	-0.50000	-2.23607	-1.72913	0.30485
11	20	100	98.50	3.5	-0.42857	-1.91663	-1.72913	0.41961
12	20	100	98.75	3.5	-0.35714	-1.59719	-1.72913	0.54173
13	20	100	99.00	3.5	-0.28571	-1.27775	-1.72913	0.66005
14	20	100	99.25	3.5	-0.21429	-0.95831	-1.72913	0.76440
15	20	100	99.50	3.5	-0.14286	-0.63888	-1.72913	0.84815
16	20	100	99.75	3.5	-0.07143	-0.31944	-1.72913	0.90933
17	20	100	100.00	3.5	0.00000	0.00000	-1.72913	0.95000
18	20	100	100.25	3.5	0.07143	0.31944	-1.72913	0.97460
19	20	100	100.50	3.5	0.14286	0.63888	-1.72913	0.98814
20	20	100	100.75	3.5	0.21429	0.95831	-1.72913	0.99492
21	20	100	101.00	3.5	0.28571	1.27775	-1.72913	0.99800

JMP Data Table and formulas:

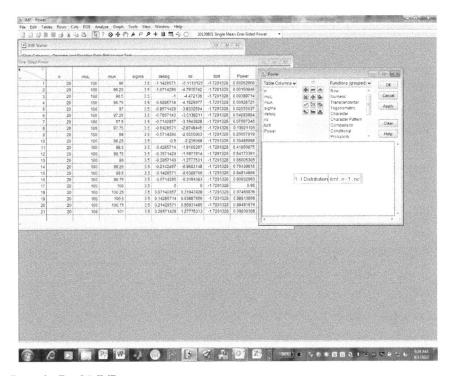

Power for Test 2.1, JMP screen.

R:

```
> muA <- c(96,96.25,96.5,96.75,97,97.25,97.5,97.75,98,98.25,
98.5,98.75,99,99.25,99.5,99.75,100,100.25,100.5,100.75,101)
> muL <- 100
> n <- 20
> sigma <- 3.5
> delta <- (muA - muL)/sigma
> nc = sqrt(n)*delta
> tcrit <- qt(0.05,n-1)
> power <- 1 - pt(tcrit,n-1,nc)
> plot(delta,power)
```

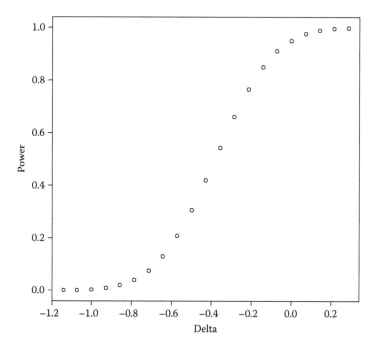

Test 2.1, R: power curve.

Test 2.2 Comparison of Two Means—Two Independent Samples, Fixed Δ Paradigm (One-Sided)

Input parameters:
    D0 = maximum desirable difference between means
    mu1 = prior estimate of mean for group 1
    n = sample size per group

SAS code:

```
libname stuff 'H:\Personal Data\Equivalence & Noninferiority\
Programs & Output';

data calc;
  set stuff.d20120803_test_2_2_two_means;
  mu2 = mu1 - Da;
  delta = (D0 - Da)/sigma;
  nc = sqrt(n/2) * delta;
  tcrit = tinv(0.05,2*n-2);
  power = 1 - probt(tcrit,2*n-2,nc);/* Pr{T' > = tcrit} */
  run;
```

```
proc print data = calc;/*dataset calc has columns n muL muA
sigma delta nc power */

 run;
```

The SAS System   13:57 Friday, August 3, 2012 2

Obs	n	mu1	D0	Da	sigma	mu2	delta	nc	tcrit	power
1	13	100	5	10.00	3.5	90.00	-1.42857	-3.64216	-1.71088	0.02925
2	13	100	5	9.75	3.5	90.25	-1.35714	-3.46005	-1.71088	0.04315
3	13	100	5	9.50	3.5	90.50	-1.28571	-3.27794	-1.71088	0.06195
4	13	100	5	9.25	3.5	90.75	-1.21429	-3.09583	-1.71088	0.08662
5	13	100	5	9.00	3.5	91.00	-1.14286	-2.91373	-1.71088	0.11797
6	13	100	5	8.75	3.5	91.25	-1.07143	-2.73162	-1.71088	0.15661
7	13	100	5	8.50	3.5	91.50	-1.00000	-2.54951	-1.71088	0.20278
8	13	100	5	8.25	3.5	91.75	-0.92857	-2.36740	-1.71088	0.25624
9	13	100	5	8.00	3.5	92.00	-0.85714	-2.18529	-1.71088	0.31626
10	13	100	5	7.75	3.5	92.25	-0.78571	-2.00319	-1.71088	0.38156
11	13	100	5	7.50	3.5	92.50	-0.71429	-1.82108	-1.71088	0.45044
12	13	100	5	7.25	3.5	92.75	-0.64286	-1.63897	-1.71088	0.52085
13	13	100	5	7.00	3.5	93.00	-0.57143	-1.45686	-1.71088	0.59063
14	13	100	5	6.75	3.5	93.25	-0.50000	-1.27475	-1.71088	0.65765
15	13	100	5	6.50	3.5	93.50	-0.42857	-1.09265	-1.71088	0.72004
16	13	100	5	6.25	3.5	93.75	-0.35714	-0.91054	-1.71088	0.77634
17	13	100	5	6.00	3.5	94.00	-0.28571	-0.72843	-1.71088	0.82558
18	13	100	5	5.75	3.5	94.25	-0.21429	-0.54632	-1.71088	0.86732
19	13	100	5	5.50	3.5	94.50	-0.14286	-0.36422	-1.71088	0.90161
20	13	100	5	5.25	3.5	94.75	-0.07143	-0.18211	-1.71088	0.92892
21	13	100	5	5.00	3.5	95.00	0.00000	0.00000	-1.71088	0.95000

JMP Data Table and formulas:

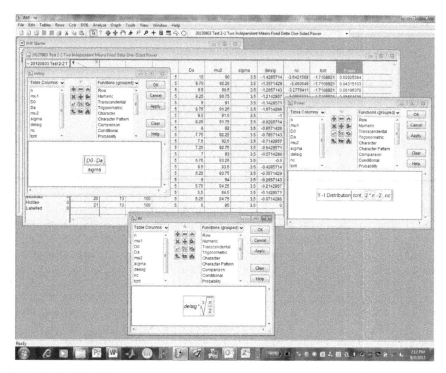

Power for Test 2.2, JMP screen.

R:

```
> df1 <- read.table("H:\\Personal Data\\Equivalence &
Noninferiority\\Programs & Output\\d20120803_test_2_2_two_
means.csv",header = TRUE,sep = ",")
> df1
      n   mu1  D0   Da   sigma
1    13   100   5  10.00   3.5
2    13   100   5   9.75   3.5
3    13   100   5   9.50   3.5
4    13   100   5   9.25   3.5
5    13   100   5   9.00   3.5
6    13   100   5   8.75   3.5
7    13   100   5   8.50   3.5
8    13   100   5   8.25   3.5
9    13   100   5   8.00   3.5
10   13   100   5   7.75   3.5
11   13   100   5   7.50   3.5
12   13   100   5   7.25   3.5
```

```
13   13   100   5   7.00   3.5
14   13   100   5   6.75   3.5
15   13   100   5   6.50   3.5
16   13   100   5   6.25   3.5
17   13   100   5   6.00   3.5
18   13   100   5   5.75   3.5
19   13   100   5   5.50   3.5
20   13   100   5   5.25   3.5
21   13   100   5   5.00   3.5
> mu2 <- df1$mu1 - df1$Da
> delsig <- (df1$D0 - df1$Da)/df1$sigma
> nc <- delsig*sqrt(df1$n/2)
> tcrit <- qt(.05,2*df1$n-2)
> power <- 1 - pt(tcrit,2*df1$n-2,nc)
> power
[1] 0.02925394 0.04315103 0.06195378 0.08661506 0.11796964
0.15661283
[7] 0.20277933 0.25624291 0.31625815 0.38156107 0.45043685
0.52085077
[13] 0.59062639 0.65764531 0.72003815 0.77633847 0.82557882
0.86731959
[19] 0.90161406 0.92892317 0.95000000
> plot(delsig,power)
```

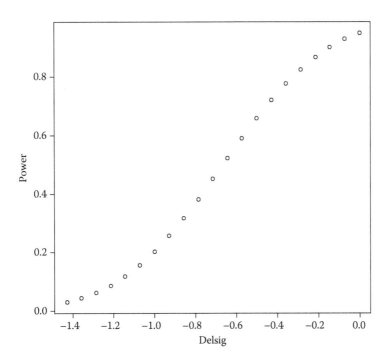

Test 2.2, R: power curve.

Test 2.3 Comparison of Two Means—Proportional Difference Paradigm for Two Independent Samples (One-Sided)

Input parameters:
mu2 = prior estimate of mean for group 2
p0 = desired maximum proportion difference between mean of group 1 relative to mean of group 2
n = sample size per group

SAS code:

```
libname stuff 'H:\Personal Data\Equivalence & Noninferiority\
Programs & Output';

data calc;
  set stuff.d20120805_test_2_3_two_means_pro;
  mu1 = (1-pa)*mu2;
  se = sigma*sqrt(1+(1-p0)**2);
  delta = (p0 - pa)*mu2/se;
  nc = sqrt(n) * delta;
  tcrit = tinv(0.05,2*n-2);
  power = 1 - probt(tcrit,2*n-2,nc);/* Pr{T' > = tcrit} */
  run;

proc print data = calc;/*dataset calc has columns n muL muA
sigma delta nc power */

  run;
```

```
              The SAS System          07:15 Sunday, August 5, 2012 1

Obs  n  mu2  p0    pa     sigma mu1    se       delta      nc        tcrit      power

 1  15  100  0.05  0.0500  2.8  95.00  3.86207   0.00000   0.00000  -1.70113  0.95000
 2  15  100  0.05  0.0550  2.8  94.50  3.86207  -0.12946  -0.50141  -1.70113  0.87605
 3  15  100  0.05  0.0600  2.8  94.00  3.86207  -0.25893  -1.00283  -1.70113  0.74736
 4  15  100  0.05  0.0650  2.8  93.50  3.86207  -0.38839  -1.50424  -1.70113  0.57027
 5  15  100  0.05  0.0700  2.8  93.00  3.86207  -0.51786  -2.00565  -1.70113  0.37753
 6  15  100  0.05  0.0750  2.8  92.50  3.86207  -0.64732  -2.50706  -1.70113  0.21160
 7  15  100  0.05  0.0800  2.8  92.00  3.86207  -0.77679  -3.00848  -1.70113  0.09858
 8  15  100  0.05  0.0835  2.8  91.65  3.86207  -0.86741  -3.35946  -1.70113  0.05136
 9  15  100  0.05  0.0900  2.8  91.00  3.86207  -1.03571  -4.01130  -1.70113  0.01170
10  15  100  0.05  0.0950  2.8  90.50  3.86207  -1.16518  -4.51271  -1.70113  0.00293
11  15  100  0.05  0.0900  2.8  91.00  3.86207  -1.03571  -4.01130  -1.70113  0.01170
12  15  100  0.05  0.1000  2.8  90.00  3.86207  -1.29464  -5.01413  -1.70113  0.00059
13  15  100  0.05  0.1050  2.8  89.50  3.86207  -1.42411  -5.51554  -1.70113  0.00010
14  15  100  0.05  0.1100  2.8  89.00  3.86207  -1.55357  -6.01695  -1.70113  0.00001
15  15  100  0.05  0.1150  2.8  88.50  3.86207  -1.68303  -6.51836  -1.70113  0.00000
16  15  100  0.05  0.1200  2.8  88.00  3.86207  -1.81250  -7.01978  -1.70113  0.00000
```

JMP Data Table and formulas:

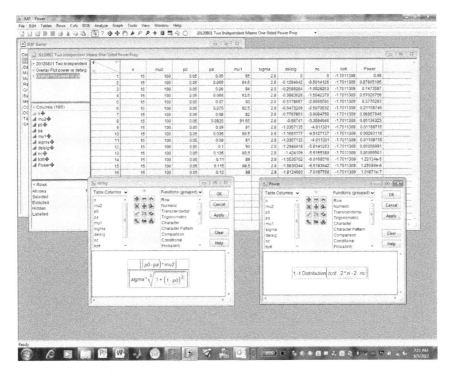

Power for Test 2.3, JMP screen.

R:

```
> df1 <- read.table("H:\\Personal Data\\Equivalence &
Noninferiority\\Programs & Output\\d20120805_test_2_3_two_
means_pro.csv",header = TRUE,sep = ",")
> df1
    n   mu2    p0      pa   sigma
1   15  100   0.05  0.0500   2.8
2   15  100   0.05  0.0550   2.8
3   15  100   0.05  0.0600   2.8
4   15  100   0.05  0.0650   2.8
5   15  100   0.05  0.0700   2.8
6   15  100   0.05  0.0750   2.8
7   15  100   0.05  0.0800   2.8
8   15  100   0.05  0.0835   2.8
9   15  100   0.05  0.0900   2.8
10  15  100   0.05  0.0950   2.8
11  15  100   0.05  0.0900   2.8
12  15  100   0.05  0.1000   2.8
13  15  100   0.05  0.1050   2.8
```

```
14   15   100   0.05   0.1100   2.8
15   15   100   0.05   0.1150   2.8
16   15   100   0.05   0.1200   2.8
> attach(df1)
> mu1 <- (1-p0)*mu2
> se <- sigma*sqrt(1 ± (1-p0)**2)
> delta <- (p0 - pa)*mu2/se
> nc <- sqrt(n)*delta
> tcr:t <- qt(0.05,2n-2)
> power <- 1 - pt(tcrit,2*n-2,nc)
> power
[1] 9.500000e-01 8.757451e-01 7.463953e-01 5.684745e-01
3.751986e-01
[6] 2.093631e-01 9.696149e-02 5.026319e-02 1.131429e-02
2.804476e-03
[11] 1.131429e-02 5.573024e-04 8.846726e-05 1.118841e-05
1.125064e-06
[16] 8.981452e-08
> plot(delta,power)
```

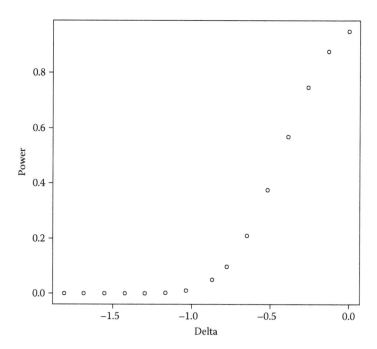

Test 2.3, R: power curve.

Test 3.1 Single Variance (One-Sided)

<u>Input parameters:</u>
  n = sample size
  k = multiple of maximum allowable σ_0 (that is, it is desired that k = 1; values
       of k > 1 are undesirable)
  beta = probability of failure to reject the null hypothesis given that k = 1

SAS code:

```
libname stuff 'H:\Personal Data\Equivalence & Noninferiority\
Programs & Output';

data calc;
  set stuff.d20120913_test_3_1_single_sd;
  chicrit = cinv(1-beta,n-1);
  chialt = chicrit/k**2;
    power = probchi(chialt,n-1);/* Pr{Chi-Sq < = chialt} */
  run;

proc print data = calc;/*dataset calc has columns n k beta
chicrit chialt power */

  run;
```

```
The SAS System     07:30 Thursday, October 25, 2012 5
```

Obs	n	k	beta	chicrit	chialt	power
1	20	0.5	0.05	30.1435	120.574	1.00000
2	20	0.6	0.05	30.1435	83.732	1.00000
3	20	0.7	0.05	30.1435	61.517	1.00000
4	20	0.8	0.05	30.1435	47.099	0.99965
5	20	0.9	0.05	30.1435	37.214	0.99254
6	20	1.0	0.05	30.1435	30.144	0.95000
7	20	1.1	0.05	30.1435	24.912	0.83654
8	20	1.2	0.05	30.1435	20.933	0.65949
9	20	1.3	0.05	30.1435	17.836	0.46661
10	20	1.4	0.05	30.1435	15.379	0.30180
11	20	1.5	0.05	30.1435	13.397	0.18236
12	20	1.6	0.05	30.1435	11.775	0.10498
13	20	1.7	0.05	30.1435	10.430	0.05848
14	20	1.8	0.05	30.1435	9.304	0.03191
15	20	1.9	0.05	30.1435	8.350	0.01721
16	20	2.0	0.05	30.1435	7.536	0.00924

JMP Data Table and formulas:

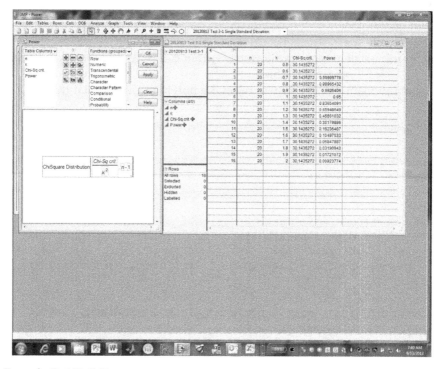

Power for Test 3.1, JMP screen.

R:

```
> df1 <- read.table("H:\\Personal Data\\Equivalence &
Noninferiority\\Programs & Output\\d20120913_test_3_1_single_
sd.csv",header = TRUE,sep = ",")
> attach(df1)
> df1
      n     k
1    20   0.5
2    20   0.6
3    20   0.7
4    20   0.8
5    20   0.9
6    20   1.0
7    20   1.1
8    20   1.2
9    20   1.3
10   20   1.4
11   20   1.5
```

```
12    20    1.6
13    20    1.7
14    20    1.8
15    20    1.9
16    20    2.0
> chicrit <- qchisq(0.95,n-1)
> chicrit
[1] 30.14353 30.14353 30.14353 30.14353 30.14353 30.14353
30.14353 30.14353
[9] 30.14353 30.14353 30.14353 30.14353 30.14353 30.14353
30.14353 30.14353
> chialt <- chicrit/(k**2)
> chialt
[1] 120.574109 83.732020 61.517402 47.099261 37.214231
30.143527
[7] 24.912006 20.933005 17.836407 15.379351 13.397123 11.774815
[13] 10.430286 9.303558 8.350008 7.535882
> power <- pchisq(chialt,n-1)
> power
[1] 1.000000000 1.000000000 0.999997779 0.999654323
0.992540600 0.950000000
[7] 0.836540906 0.659485488 0.466610324 0.301798876
0.182364875 0.104976327
[13] 0.058478875 0.031909430 0.017210722 0.009237735
> plot(k,power)
```

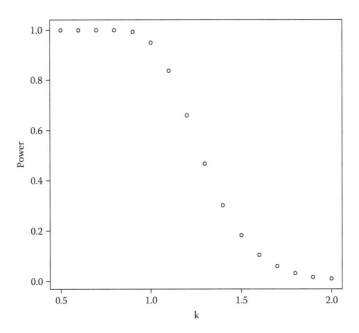

Test 3.1, R: power curve.

Test 3.2 Comparison of Two Variances (One-Sided)

Input parameters:
    nnum = sample size for numerator variance
    ndenom = sample size for denominator variance
    k0 = multiple of maximum allowable σ_0
    ka = alternative multiples (it is desired that ka = k0; values of ka > k0 are
        undesirable)

SAS code:

```
libname stuff 'H:\Personal Data\Equivalence & Noninferiority\
Programs & Output';

data calc;
  set stuff.d20120914_test_3_2_two_sds;
  beta = 0.05;
  fcrit = finv(1-beta,nnum-1,ndenom-1);
  falt = (k0**2)*fcrit/ka**2;
    power = probf(falt,nnum-1,ndenom-1);/* Pr{F < = falt} */
  run;

proc print data = calc;/*dataset calc has columns nnum ndenom
k0 ka fcrit falt power */

  run;
```

The SAS System   07:30 Thursday, October 25, 2012 6

Obs	nnum	ndenom	k0	ka	beta	fcrit	falt	power
1	20	20	1.1	0.5	0.05	2.16825	10.4943	1.00000
2	20	20	1.1	0.6	0.05	2.16825	7.2877	0.99997
3	20	20	1.1	0.7	0.05	2.16825	5.3543	0.99970
4	20	20	1.1	0.8	0.05	2.16825	4.0994	0.99827
5	20	20	1.1	0.9	0.05	2.16825	3.2390	0.99308
6	20	20	1.1	1.0	0.05	2.16825	2.6236	0.97917
7	20	20	1.1	1.1	0.05	2.16825	2.1683	0.95000
8	20	20	1.1	1.2	0.05	2.16825	1.8219	0.89990
9	20	20	1.1	1.3	0.05	2.16825	1.5524	0.82698
10	20	20	1.1	1.4	0.05	2.16825	1.3386	0.73438
11	20	20	1.1	1.5	0.05	2.16825	1.1660	0.62940
12	20	20	1.1	1.6	0.05	2.16825	1.0248	0.52104
13	20	20	1.1	1.7	0.05	2.16825	0.9078	0.41762
14	20	20	1.1	1.8	0.05	2.16825	0.8097	0.32509
15	20	20	1.1	1.9	0.05	2.16825	0.7268	0.24664
16	20	20	1.1	2.0	0.05	2.16825	0.6559	0.18303

17	20	20	1.1	2.1	0.05	2.16825	0.5949	0.13329
18	20	20	1.1	2.2	0.05	2.16825	0.5421	0.09556

JMP Data Table and formulas:

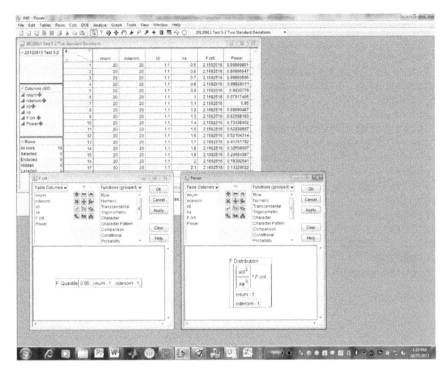

Power for Test 3.2, JMP screen.

R:

```
> df1 <- read.table("H:\\Personal Data\\Equivalence &
Noninferiority\\Programs & Output\\d20120914_test_3_2_two_sds.
csv",header = TRUE,sep = ",")
> df1
    nnum ndenom  k0    ka
1    20     20   1.1   0.5
2    20     20   1.1   0.6
3    20     20   1.1   0.7
4    20     20   1.1   0.8
5    20     20   1.1   0.9
6    20     20   1.1   1.0
7    20     20   1.1   1.1
8    20     20   1.1   1.2
9    20     20   1.1   1.3
10   20     20   1.1   1.4
```

```
11    20    20    1.1    1.5
12    20    20    1.1    1.6
13    20    20    1.1    1.7
14    20    20    1.1    1.8
15    20    20    1.1    1.9
16    20    20    1.1    2.0
17    20    20    1.1    2.1
18    20    20    1.1    2.2
> beta <- 0.05
> fcrit <- qf(1-beta,nnum-1,ndenom-1)
Error in qf(1 - beta, nnum - 1, ndenom - 1) : object 'nnum'
not found
> attach(df1)
> fcrit <- qf(1-beta,nnum-1,ndenom-1)
> fac <- k0**2/ka**2
> power <- pf(fac*fcrit,nnum-1,ndenom-1)
> power
[1] 0.99999801 0.99996647 0.99969596 0.99826511 0.99307790
0.97917405
[7] 0.95000000 0.89990487 0.82698183 0.73438402 0.62939587
0.52104314
[13] 0.41761782 0.32509097 0.24664387 0.18302541 0.13329022
0.09556305
> plot(ka,power)
```

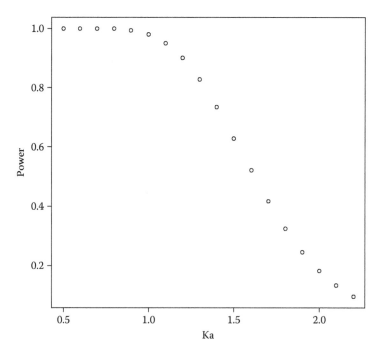

Test 3.2, R: power curve.

Test 3.3 Single Coefficient of Variation (One-Sided)

Input parameters:
  n = sample size
  C0 = maximum desirable coefficient of variation (percent)
  cv_pct = alternative CV values (percent)
  beta = probability of failure to reject the null hypothesis given that
        cv_pct = C0

SAS code:

```
libname stuff 'H:\Personal Data\Equivalence & Noninferiority\
Programs & Output';

data calc;
  set stuff.d20121016_test_3_3_cv_power_curv;/* input C0 n
  cv_pct beta */
  nct_null = sqrt(n)/(C0/100);
  tcrit = tinv(beta,n-1,nct_null);
  cv_crit = 100*(sqrt(n)/tcrit);
  nct = sqrt(n)/(cv_pct/100);
  t_alt = tinv(beta,n-1,nct);
  cv_alt = 100*(sqrt(n)/t_alt);
  power = 1 - probt(tcrit,n-1,nct);/* Pr{t = > tcrit|nct} */
  run;

proc print data = calc;/*dataset calc has columns C0 n cv_pct
beta nct_null tcrtit cv-crit t_alt cv_alt power */

  run;
```

```
          The SAS System        06:08 Sunday, October 21, 2012 3

Obs C0  n cv_pct beta nct_null  tcrit  cv_crit   nct   t_alt  cv_alt  power

  1  6 64   5.75 0.05 133.333 116.428 6.87122 139.130 121.495  6.5846 0.98516
  2  6 64   6.00 0.05 133.333 116.428 6.87122 133.333 116.428  6.8712 0.95000
  3  6 64   6.25 0.05 133.333 116.428 6.87122 128.000 111.765  7.1579 0.87528
  4  6 64   6.50 0.05 133.333 116.428 6.87122 123.077 107.461  7.4446 0.75517
  5  6 64   6.75 0.05 133.333 116.428 6.87122 118.519 103.475  7.7313 0.60216
  6  6 64   7.00 0.05 133.333 116.428 6.87122 114.286  99.774  8.0181 0.44157
  7  6 64   7.25 0.05 133.333 116.428 6.87122 110.345  96.328  8.3049 0.29826
  8  6 64   7.50 0.05 133.333 116.428 6.87122 106.667  93.112  8.5918 0.18659
  9  6 64   7.75 0.05 133.333 116.428 6.87122 103.226  90.103  8.8787 0.10893
 10  6 64   8.00 0.05 133.333 116.428 6.87122 100.000  87.282  9.1657 0.05982
 11  6 64   8.25 0.05 133.333 116.428 6.87122  96.970  84.631  9.4528 0.03115
 12  6 64   8.50 0.05 133.333 116.428 6.87122  94.118  82.137  9.7399 0.01550
 13  6 64   8.75 0.05 133.333 116.428 6.87122  91.429  79.784 10.0270 0.00742
 14  6 64   9.00 0.05 133.333 116.428 6.87122  88.889  77.563 10.3142 0.00344
 15  6 64   9.25 0.05 133.333 116.428 6.87122  86.486  75.461 10.6015 0.00155
 16  6 64   9.50 0.05 133.333 116.428 6.87122  84.211  73.470 10.8889 0.00068
 17  6 64   9.75 0.05 133.333 116.428 6.87122  82.051  71.580 11.1763 0.00030
 18  6 64  10.00 0.05 133.333 116.428 6.87122  80.000  69.785 11.4637 0.00013
```

JMP Data Table and formulas:

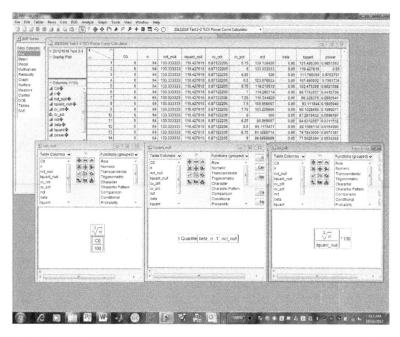

Power for Test 3.3, JMP screen 1.

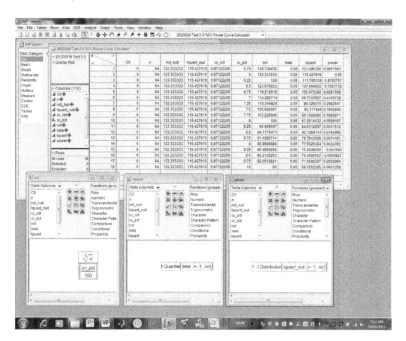

Power for Test 3.3, JMP screen 2.

R:

```
df1 <- read.table("H:\\Personal Data\\Equivalence &
Noninferiority\\Programs & Output\\d20121021_test_3_3_cv_
power.csv",header = TRUE,sep = ",")
> df1
      C0    n    beta    cv_pct
 1     6   64    0.05     5.75
 2     6   64    0.05     6.00
 3     6   64    0.05     6.25
 4     6   64    0.05     6.50
 5     6   64    0.05     6.75
 6     6   64    0.05     7.00
 7     6   64    0.05     7.25
 8     6   64    0.05     7.50
 9     6   64    0.05     7.75
10     6   64    0.05     8.00
11     6   64    0.05     8.25
12     6   64    0.05     8.50
13     6   64    0.05     8.75
14     6   64    0.05     9.00
15     6   64    0.05     9.25
16     6   64    0.05     9.50
17     6   64    0.05     9.75
18     6   64    0.05    10.00
> attach(df1)
> nct_null <- sqrt(n)/(C0/100)
> tcrit <- qt(beta,n-1,nct_null)
> nct <- sqrt(n)/(cv_pct/100)
> t_alt <- qt(beta,n-1,nct)
> cv_alt <- 100*sqrt(n)/t_alt
> power <- 1 - pt(tcrit,n-1,nct)
> power
 [1] 0.9861050152 0.9500000000 0.8715816329 0.7460943038
0.5892190044
 [6] 0.4285830459 0.2885809310 0.1814330537 0.1075681294
0.0607377413
[11] 0.0329631123 0.0173362572 0.0088991809 0.0044862788
0.0022327487
[16] 0.0011018783 0.0005412315 0.0002654232
> plot(cv_alt,power)
```

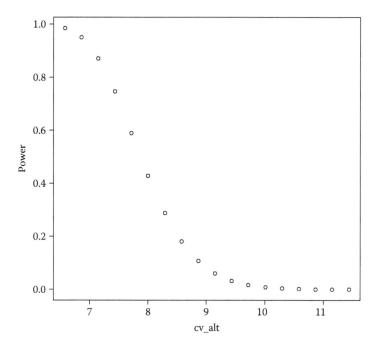

Test 3.3, R: power curve.

NOTE: the calculation of the critical t percentile using R resulted in a value of 116.627. Both SAS and JMP calculated a value of 116.428. Thus, power calculations were slightly different for R in this example.

Test 4.1 Single Exponential Rate Parameter (One-Sided)

Input parameters:
  n = sample size
  lamda_0 = maximum desirable (failure) rate
  beta = probability of failure to reject the null hypothesis given that lamda_a
      = lamda_0

SAS code:

```
libname stuff 'H:\Personal Data\Equivalence & Noninferiority\
Programs & Output';

data calc;
   set stuff.d20121022_test_4_1_exponential;/* input n lamda_0
   lamda_a */
```

```
beta = 0.05;
k0 = gaminv(beta,n)/lamda_0;/* note that SAS gaminv and
probgam assume lamda = scale parameter = 1 */
power = 1-probgam(k0*lamda_a,n);/* Pr{gamma > = k0|n,
lamda_a} */
run;

proc print data = calc;/*dataset calc has columns n lamda_0
lamda_a beta k0 ka power */

run;
```

The SAS System   08:52 Thursday, November 8, 2012 18

Obs	n	lamda_0	lamda_a	beta	k0	power
1	20	0.75	0.75	0.05	17.6729	0.95000
2	20	0.75	0.80	0.05	17.6729	0.91771
3	20	0.75	0.85	0.05	17.6729	0.87399
4	20	0.75	0.90	0.05	17.6729	0.81879
5	20	0.75	0.95	0.05	17.6729	0.75325
6	20	0.75	1.00	0.05	17.6729	0.67963
7	20	0.75	1.05	0.05	17.6729	0.60095
8	20	0.75	1.10	0.05	17.6729	0.52057
9	20	0.75	1.15	0.05	17.6729	0.44175
10	20	0.75	1.20	0.05	17.6729	0.36730
11	20	0.75	1.25	0.05	17.6729	0.29937
12	20	0.75	1.30	0.05	17.6729	0.23931
13	20	0.75	1.35	0.05	17.6729	0.18773
14	20	0.75	1.40	0.05	17.6729	0.14462
15	20	0.75	1.45	0.05	17.6729	0.10949
16	20	0.75	1.50	0.05	17.6729	0.08152
17	20	0.75	1.55	0.05	17.6729	0.05973
18	20	0.75	1.60	0.05	17.6729	0.04310
19	20	0.75	1.65	0.05	17.6729	0.03064
20	20	0.75	1.70	0.05	17.6729	0.02148

JMP Data Table and formulas:

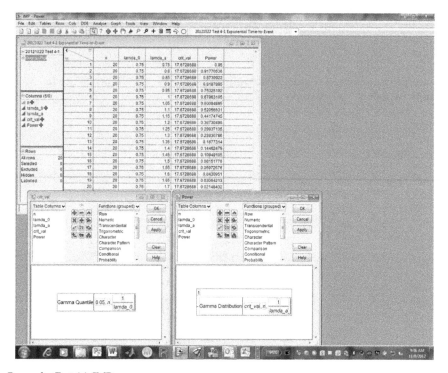

Power for Test 4.1, JMP screen.

R:

```
> df1 <- read.table("H:\\Personal Data\\Equivalence &
Noninferiority\\Programs & Output\\d20121022_test_4_1_exponen-
tial.csv",header = TRUE,sep = ",")
> attach(df1)
> df1
       n  lamda_0  lamda_a
 1    20    0.75     0.75
 2    20    0.75     0.80
 3    20    0.75     0.85
 4    20    0.75     0.90
 5    20    0.75     0.95
 6    20    0.75     1.00
 7    20    0.75     1.05
 8    20    0.75     1.10
 9    20    0.75     1.15
10    20    0.75     1.20
11    20    0.75     1.25
12    20    0.75     1.30
```

```
13    20    0.75    1.35
14    20    0.75    1.40
15    20    0.75    1.45
16    20    0.75    1.50
17    20    0.75    1.55
18    20    0.75    1.60
19    20    0.75    1.65
20    20    0.75    1.70
> scale0 <- 1/lamda_0
> crit_gam <- qgamma(0.05,shape = n,scale = scale0)
> crit_gam
[1] 17.67287 17.67287 17.67287 17.67287 17.67287 17.67287
17.67287 17.67287
[9] 17.67287 17.67287 17.67287 17.67287 17.67287 17.67287
17.67287 17.67287
[17] 17.67287 17.67287 17.67287 17.67287
> scalea <- 1/lamda_a
> scalea
[1] 1.3333333 1.2500000 1.1764706 1.1111111 1.0526316
1.0000000 0.9523810
[8] 0.9090909 0.8695652 0.8333333 0.8000000 0.7692308
0.7407407 0.7142857
[15] 0.6896552 0.6666667 0.6451613 0.6250000 0.6060606
0.5882353
> power <- 1 - pgamma(crit_gam,shape = n,scale = scalea)
> power
[1] 0.95000000 0.91770536 0.87399220 0.81878950 0.75325192
0.67963105
[7] 0.60094895 0.52056631 0.44174745 0.36730496 0.29937135
0.23930786
[13] 0.18773140 0.14462479 0.10949105 0.08151776 0.05972676
0.04309510
[19] 0.03064213 0.02148432
> plot(lamda_a,power)
```

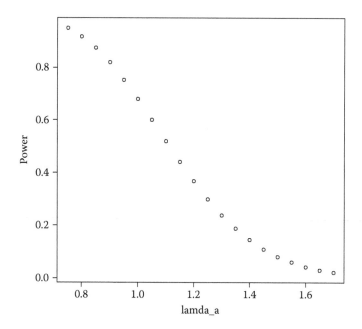

Test 4.1, R: power curve.

Test 4.2 Two Exponential Rate Parameters (One-Sided)

Input parameters:
    n = sample size
    del0 = maximum desirable multiple of nominal rate
    dela = alternative multipliers; dela > del0
    beta = probability of failure to reject the null hypothesis given del0

SAS code:

```
libname stuff 'H:\Personal Data\Equivalence & Noninferiority\
Programs & Output';

data calc;
  set stuff.d20121024_test_4_2_two_exp;

  delta = (1/dela) - (1/del0);
  nc = sqrt(n) * delta/sqrt((1 + (1/del0)**2));
  tcrit = tinv(beta,2*n-2);
  power = 1 - probt(tcrit,2*n-2,nc);/* Pr{T' > = tcrit} */
  run;
```

```
proc print data = calc;/*dataset calc has columns n beta del0
dela delta nc tcrit power */

  run;
```

          The SAS System   07:36 Wednesday, October 24, 2012 2

Obs	n	beta	del0	dela	delta	nc	tcrit	power
1	30	0.05	1	1.0	0.00000	0.00000	-1.67155	0.95000
2	30	0.05	1	1.1	-0.09091	-0.35209	-1.67155	0.90266
3	30	0.05	1	1.2	-0.16667	-0.64550	-1.67155	0.84300
4	30	0.05	1	1.3	-0.23077	-0.89377	-1.67155	0.77682
5	30	0.05	1	1.4	-0.28571	-1.10657	-1.67155	0.70925
6	30	0.05	1	1.5	-0.33333	-1.29099	-1.67155	0.64391
7	30	0.05	1	1.6	-0.37500	-1.45237	-1.67155	0.58295
8	30	0.05	1	1.7	-0.41176	-1.59476	-1.67155	0.52740
9	30	0.05	1	1.8	-0.44444	-1.72133	-1.67155	0.47753
10	30	0.05	1	1.9	-0.47368	-1.83457	-1.67155	0.43319
11	30	0.05	1	2.0	-0.50000	-1.93649	-1.67155	0.39397
12	30	0.05	1	2.1	-0.52381	-2.02871	-1.67155	0.35939
13	30	0.05	1	2.2	-0.54545	-2.11254	-1.67155	0.32891
14	30	0.05	1	2.3	-0.56522	-2.18908	-1.67155	0.30203
15	30	0.05	1	2.4	-0.58333	-2.25924	-1.67155	0.27830
16	30	0.05	1	2.5	-0.60000	-2.32379	-1.67155	0.25730
17	30	0.05	1	2.6	-0.61538	-2.38337	-1.67155	0.23868
18	30	0.05	1	2.7	-0.62963	-2.43855	-1.67155	0.22211
19	30	0.05	1	2.8	-0.64286	-2.48978	-1.67155	0.20733
20	30	0.05	1	2.9	-0.65517	-2.53747	-1.67155	0.19411
21	30	0.05	1	3.0	-0.66667	-2.58199	-1.67155	0.18225
22	30	0.05	1	4.0	-0.75000	-2.90474	-1.67155	0.11015
23	30	0.05	1	5.0	-0.80000	-3.09839	-1.67155	0.07823
24	30	0.05	1	6.0	-0.83333	-3.22749	-1.67155	0.06122
25	30	0.05	1	7.0	-0.85714	-3.31970	-1.67155	0.05095
26	30	0.05	1	8.0	-0.87500	-3.38886	-1.67155	0.04418
27	30	0.05	1	9.0	-0.88889	-3.44265	-1.67155	0.03944
28	30	0.05	1	10.0	-0.90000	-3.48569	-1.67155	0.03595

JMP Data Table and formulas:

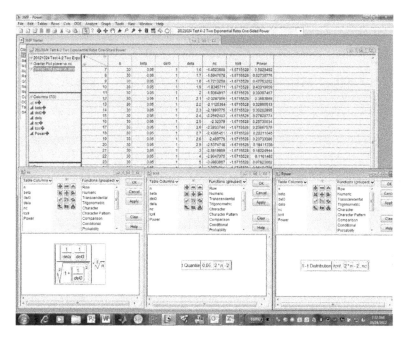

Power for Test 4.2, JMP screen.

R:

```
> df1 <- read.table("H:\\Personal Data\\Equivalence &
Noninferiority\\Programs & Output\\d20121024_test_4_2_two_exp.
csv",header = TRUE,sep = ",")
> df1
        n    beta   del0    dela
1      30    0.05    1     1.0
2      30    0.05    1     1.1
3      30    0.05    1     1.2
4      30    0.05    1     1.3
5      30    0.05    1     1.4
6      30    0.05    1     1.5
7      30    0.05    1     1.6
8      30    0.05    1     1.7
9      30    0.05    1     1.8
10     30    0.05    1     1.9
11     30    0.05    1     2.0
12     30    0.05    1     2.1
13     30    0.05    1     2.2
14     30    0.05    1     2.3
15     30    0.05    1     2.4
```

```
16    30    0.05    1    2.5
17    30    0.05    1    2.6
18    30    0.05    1    2.7
19    30    0.05    1    2.8
20    30    0.05    1    2.9
21    30    0.05    1    3.0
22    30    0.05    1    4.0
23    30    0.05    1    5.0
24    30    0.05    1    6.0
25    30    0.05    1    7.0
26    30    0.05    1    8.0
27    30    0.05    1    9.0
28    30    0.05    1    10.0
> attach(df1)
> delta <- (1/dela) - (1/del0)
> nc <- sqrt(n)*delta/sqrt(1±(1/del0)**2)
> tcrit <- qt(beta,2*n-2)
> power <- 1 - pt(tcrit,2*n-2,nc)
> power
[1]  0.95000000 0.90265934 0.84300024 0.77681996 0.70924735
0.64390729
[7]  0.58294920 0.52739776 0.47753262 0.43319059 0.39397467
0.35938660
[13] 0.32890513 0.30202895 0.27829774 0.25730034 0.23867575
0.22211045
[19] 0.20733386 0.19411338 0.18224944 0.11014930 0.07823062
0.06121769
[25] 0.05094668 0.04418344 0.03943986 0.03595075
> plot(dela,power)
```

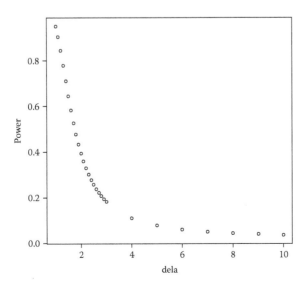

Test 4.2, R: power curve.

Test 5.1 C_{pk}

Input parameters:
  n = sample size
  cpk0 = minimum desirable C_{pk}
  cpka = alternative C_{pk} values; cpka ≤ cpk0
  beta = probability of failure to reject the null hypothesis given cpk0

SAS code:

```
libname stuff 'H:\Personal Data\Equivalence & Noninferiority\
Programs & Output';

data calc;
  set stuff.d20121023_test_5_1_power_cpk;
  nc0 = 3*sqrt(n)*cpk0;
  nca = 3*sqrt(n)*cpka;
  tcrit = tinv(beta,n-1,nc0);
  cpkcrit = tcrit/(3*sqrt(n));
  bias = sqrt((n-1)/2)*gamma((n-2)/2)/gamma((n-1)/2);
  cpk_bcorr = cpkcrit*bias;
  power = 1 - probt(tcrit,n-1,nca);/* Pr{T' > = tcrit} */
  run;

proc print data = calc;/*dataset calc has columns cpk0 cpka n
beta tcrit cpkcrit bias cpk_bcorr power */

  run;
```

```
           The SAS System        11:43 Tuesday, October 23, 2012 2
```

Obs	Cpk0	Cpka	n	beta	nc0	nca	tcrit	cpkcrit	bias	cpk_bcorr	power
1	1.33	1.33	100	0.05	39.9	39.9	35.4878	1.18293	1.00766	1.19198	0.95000
2	1.33	1.30	100	0.05	39.9	39.0	35.4878	1.18293	1.00766	1.19198	0.90687
3	1.33	1.25	100	0.05	39.9	37.5	35.4878	1.18293	1.00766	1.19198	0.78202
4	1.33	1.22	100	0.05	39.9	36.6	35.4878	1.18293	1.00766	1.19198	0.67396
5	1.33	1.20	100	0.05	39.9	36.0	35.4878	1.18293	1.00766	1.19198	0.59134
6	1.33	1.15	100	0.05	39.9	34.5	35.4878	1.18293	1.00766	1.19198	0.37345
7	1.33	1.10	100	0.05	39.9	33.0	35.4878	1.18293	1.00766	1.19198	0.18874
8	1.33	1.05	100	0.05	39.9	31.5	35.4878	1.18293	1.00766	1.19198	0.07369
9	1.33	1.02	100	0.05	39.9	30.6	35.4878	1.18293	1.00766	1.19198	0.03657
10	1.33	1.00	100	0.05	39.9	30.0	35.4878	1.18293	1.00766	1.19198	0.02158
11	1.33	0.99	100	0.05	39.9	29.7	35.4878	1.18293	1.00766	1.19198	0.01627
12	1.33	0.95	100	0.05	39.9	28.5	35.4878	1.18293	1.00766	1.19198	0.00462

JMP Data Table and formulas:

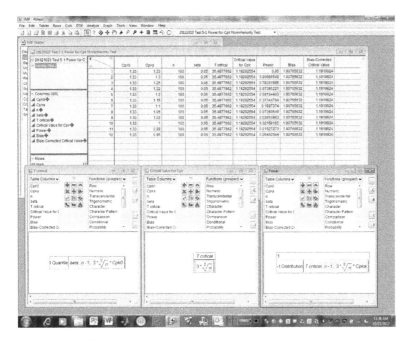

Power for Test 5.1, JMP screen 1.

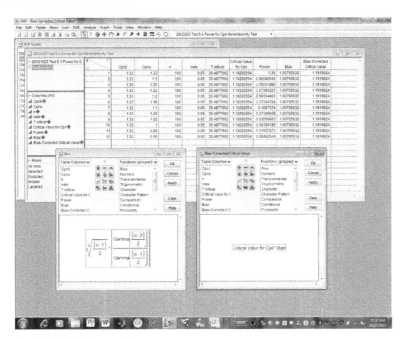

Power for Test 5.1, JMP screen 2.

R:

```
> df1 <- read.table("H:\\Personal Data\\Equivalence &
Noninferiority\\Programs & Output\\d20121023_test_5_1_cpk_
power.csv",header = TRUE,sep = ",")
> df1
     Cpk0    Cpka      n     beta
1    1.33    1.33    100     0.05
2    1.33    1.30    100     0.05
3    1.33    1.25    100     0.05
4    1.33    1.22    100     0.05
5    1.33    1.20    100     0.05
6    1.33    1.15    100     0.05
7    1.33    1.10    100     0.05
8    1.33    1.05    100     0.05
9    1.33    1.02    100     0.05
10   1.33    1.00    100     0.05
11   1.33    0.99    100     0.05
12   1.33    0.95    100     0.05
> attach(df1)
> nc0 <- 3*sqrt(n)*Cpk0
> tcrit <- qt(beta,n-1,nc0)
> cpk_crit <- tcrit/(3*sqrt(n))
> nca <- 3*sqrt(n)*Cpka
> power = 1 - pt(tcrit,n-1,nca)
> bias <- sqrt((n-1)/2)*gamma((n-2)/2)/gamma((n-1)/2)
> cpk_bcorr <- cpk_crit*bias
> power
[1] 0.950000000 0.905477195 0.778011372 0.669142561
0.586209919 0.368588986
[7] 0.185372972 0.072000352 0.035620438 0.020977501
0.015800452 0.004470561
> plot(Cpka,power)
```

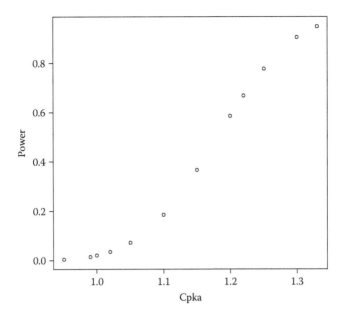

Test 5.1, R: power curve.

Test 5.2 C_p

Input parameters:
  n = sample size
  cp0 = minimum desirable C_p
  k = multiplier to compute cpa, alternative values of C_p, cpa ≤ cp0
  beta probability of failure to reject the null hypothesis given k = 1
    (cpa = cp0)

SAS code:

```
libname stuff 'H:\Personal Data\Equivalence & Noninferiority\
Programs & Output';

data calc;
  set stuff.d20121023_test_5_2_cp_power;
  k = cp0/cpa;
  chicrit = cinv(1-beta,n-1);
  chialt = chicrit/k**2;
  power = probchi(chialt,n-1);/* Pr{Chi-Sq < = chialt} */
  run;
```

```
proc print data = calc;/*dataset calc has columns n cp0 cpa k
chicrit chialt power */

   run;
```

The SAS System 15:24 Tuesday, October 23, 2012 1

Obs	n	beta	cp0	cpa	k	chicrit	chialt	power
1	20	0.05	1.33	1.33	1.00000	30.1435	30.1435	0.95000
2	20	0.05	1.33	1.32	1.00758	30.1435	29.6919	0.94415
3	20	0.05	1.33	1.31	1.01527	30.1435	29.2438	0.93775
4	20	0.05	1.33	1.30	1.02308	30.1435	28.7990	0.93079
5	20	0.05	1.33	1.29	1.03101	30.1435	28.3576	0.92322
6	20	0.05	1.33	1.28	1.03906	30.1435	27.9197	0.91501
7	20	0.05	1.33	1.27	1.04724	30.1435	27.4852	0.90615
8	20	0.05	1.33	1.26	1.05556	30.1435	27.0540	0.89660
9	20	0.05	1.33	1.25	1.06400	30.1435	26.6263	0.88634
10	20	0.05	1.33	1.24	1.07258	30.1435	26.2020	0.87534
11	20	0.05	1.33	1.23	1.08130	30.1435	25.7811	0.86359
12	20	0.05	1.33	1.22	1.09016	30.1435	25.3636	0.85108
13	20	0.05	1.33	1.21	1.09917	30.1435	24.9495	0.83779
14	20	0.05	1.33	1.20	1.10833	30.1435	24.5388	0.82371
15	20	0.05	1.33	1.15	1.15652	30.1435	22.5365	0.74163
16	20	0.05	1.33	1.10	1.20909	30.1435	20.6194	0.64179
17	20	0.05	1.33	1.00	1.33000	30.1435	17.0408	0.41290
18	20	0.05	1.33	0.95	1.40000	30.1435	15.3794	0.30180
19	20	0.05	1.33	0.90	1.47778	30.1435	13.8031	0.20495
20	20	0.05	1.33	0.80	1.66250	30.1435	10.9061	0.07304
21	20	0.05	1.33	0.70	1.90000	30.1435	8.3500	0.01721

JMP Data Table and formulas:

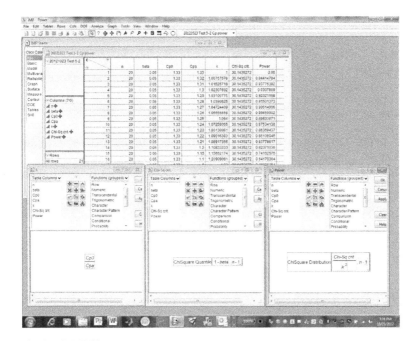

Power for Test 5.2, JMP screen.

R:

```
> df1 <- read.table("H:\\Personal Data\\Equivalence &
Noninferiority\\Programs & Output\\d20121023_test_5_2_cp_
power.csv",header = TRUE,sep = ",")
> df1
        n     beta    Cp0     Cpa
  1    20     0.05    1.33    1.33
  2
  3    20     0.05    1.33    1.31
  4    20     0.05    1.33    1.30
  5    20     0.05    1.33    1.29
  6    20     0.05    1.33    1.28
  7    20     0.05    1.33    1.27
  8    20     0.05    1.33    1.26
  9    20     0.05    1.33    1.25
 10    20     0.05    1.33    1.24
 11    20     0.05    1.33    1.23
 12    20     0.05    1.33    1.22
 13    20     0.05    1.33    1.21
 14    20     0.05    1.33    1.20
 15    20     0.05    1.33    1.15
```

```
16   20    0.05    1.33    1.10
17   20    0.05    1.33    1.00
18   20    0.05    1.33    0.95
19   20    0.05    1.33    0.90
20   20    0.05    1.33    0.80
21   20    0.05    1.33    0.70
> attach(df1)
> k <- Cp0/Cpa
> chicrit <- qchisq(1-beta,n-1)
> power <- pchisq(chicrit/(k**2),n-1)
> power
[1] 0.95000000 0.94414784 0.93775382 0.93078690 0.92321668
0.91501373
[7] 0.90614996 0.89659902 0.88633671 0.87534138 0.86359437
0.85108045
[13] 0.83778817 0.82371035 0.74162575 0.64179364 0.41289854
0.30179888
[19] 0.20495238 0.07304450 0.01721072
> plot(Cpa,power)
```

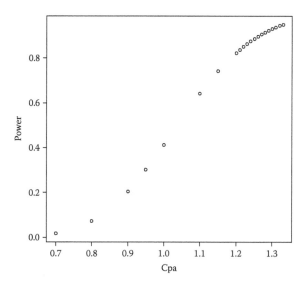

Test 5.2, R: power curve.

Test 6.1 Multivariate—Single Mean Vector

Inasmuch as inverting covariance matrices is much simpler with R compared with either JMP or SAS, only R code will be presented for this test. Also, since this test is multivariate, power is a hyper-surface over all dimensions of

the vector space of interest. Here some power calculations will be illustrated, using three-dimensional "delta" vectors (a "null" vector, delta0, and several alternatives).
Input parameters:
  n = sample size
  covmat = covariance matrix
  delta0 = vector of allowable limits on the difference between the mean vector and target mean vector
  beta = probability of failure to reject the null hypothesis given delta_alt = delta0

R:

```
> beta <- 0.05

> covmat <- c(4.5,1.6,4.8,1.6,0.6,1.65,4.8,1.65,6.3)

> dim(covmat) <-c(3,3)

> covmat

       [,1] [,2] [,3]
[1,]   4.5 1.60 4.80
[2,]   1.6 0.60 1.65
[3,]   4.8 1.65 6.30

> delta0 <- c(3,2,3)

> n <- 30

> covinv <- solve(covmat)

> covinv
            [,1]          [,2]          [,3]
[1,]    7.014925  -14.328358   -1.5920398
[2,]  -14.328358   35.223881    1.6915423
[3,]   -1.592040    1.691542    0.9286899

> nc0 <- n*t(delta0)%*%covinv%*%delta0
> nc0
          [,1]
[1,]  962.6866

> k <- 3

> F_crit <- qf(1-beta,k,n-k,nc0)

> F_crit
```

```
[1] 542.9681

> delta_alt <- c(3.3,2.2,3.3)
> nca <- n*t(delta_alt)%*%covinv%*%delta_alt
> nca
        [,1]
[1,] 1164.851
> power <- pf(F_crit,k,n-k,nca)
> power
[1] 0.8530768

> delta_alt <- c(3.6,2.4,3.6)
> nca <- n*t(delta_alt)%*%covinv%*%delta_alt
> nca
        [,1]
[1,] 1386.269
> power <- pf(F_crit,k,n-k,nca)
> power
[1] 0.6817987

> delta_alt <- c(3.9,2.8,3.9)
> nca <- n*t(delta_alt)%*%covinv%*%delta_alt
> nca
        [,1]
[1,] 2176.791
> power <- pf(F_crit,k,n-k,nca)
> power
[1] 0.1163759

> delta_alt <- c(4.0,3.0,4.0)
> nca <- n*t(delta_alt)%*%covinv%*%delta_alt
> nca
        [,1]
[1,] 2696.517
> power <- pf(F_crit,k,n-k,nca)
> power
[1] 0.01867881
```

Test 9.1 Two Sequences, Variability Known

Input parameters:
   n = sample size (number of points in each sequence)
   del0 = maximum desirable difference between the sequences
   dela = alternative differences between the sequences
   beta = probability of failure to reject the null hypothesis given del0

sigma_e = standard deviation of noise in "e" sequence
sigma_r = standard deviation of noise in "r" sequence

SAS code:

```
libname stuff 'H:\Personal Data\Equivalence & Noninferiority\
Programs & Output';

data calc;
  set stuff.d20121025_test_9_1_two_seq_var_k;
  delsig0 = del0/sqrt(sigma_e**2 + sigma_r**2);
  nc0 = n * delsig0**2;
  chi_crit = cinv(1-beta,n,nc0);
  delsiga = dela/sqrt(sigma_e**2 + sigma_r**2);
  nca = n * delsiga**2;
  power = probchi(chi_crit,n,nca);/* Pr{ChiSq < = chi_crit} */
  run;



proc print data = calc;/*dataset calc has columns n beta
sigma_e sigma_r del0 dela delsig0 delsiga chi_crit nc0 nca
power */

  run;
```

The SAS System 07:30 Thursday, October 25, 2012 1

Obs	n	sigma_e	sigma_r	beta	del0	dela	delsig0	nc0	chi_crit	delsiga	nca	power
1	30	1.665	1.665	0.05	0.33	0.33	0.14015	0.58924	44.6297	0.14015	0.5892	0.95000
2	30	1.665	1.665	0.05	0.33	0.40	0.14015	0.58924	44.6297	0.16988	0.8657	0.94582
3	30	1.665	1.665	0.05	0.33	0.50	0.14015	0.58924	44.6297	0.21234	1.3527	0.93796
4	30	1.665	1.665	0.05	0.33	0.60	0.14015	0.58924	44.6297	0.25481	1.9479	0.92747
5	30	1.665	1.665	0.05	0.33	0.70	0.14015	0.58924	44.6297	0.29728	2.6513	0.91383
6	30	1.665	1.665	0.05	0.33	0.80	0.14015	0.58924	44.6297	0.33975	3.4629	0.89641
7	30	1.665	1.665	0.05	0.33	0.90	0.14015	0.58924	44.6297	0.38222	4.3828	0.87454
8	30	1.665	1.665	0.05	0.33	1.00	0.14015	0.58924	44.6297	0.42469	5.4108	0.84756
9	30	1.665	1.665	0.05	0.33	1.50	0.14015	0.58924	44.6297	0.63703	12.1743	0.62140
10	30	1.665	1.665	0.05	0.33	2.00	0.14015	0.58924	44.6297	0.84938	21.6433	0.29742
11	30	1.665	1.665	0.05	0.33	2.50	0.14015	0.58924	44.6297	1.06172	33.8176	0.07342
12	30	1.665	1.665	0.05	0.33	2.60	0.14015	0.58924	44.6297	1.10419	36.5771	0.05036
13	30	1.665	1.665	0.05	0.33	2.70	0.14015	0.58924	44.6297	1.14666	39.4449	0.03333
14	30	1.665	1.665	0.05	0.33	2.80	0.14015	0.58924	44.6297	1.18913	42.4208	0.02127
15	30	1.665	1.665	0.05	0.33	2.90	0.14015	0.58924	44.6297	1.23160	45.5050	0.01307
16	30	1.665	1.665	0.05	0.33	3.00	0.14015	0.58924	44.6297	1.27407	48.6973	0.00773

JMP Data Table and formulas:

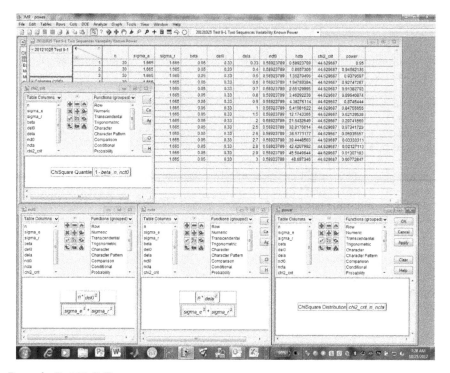

Power for Test 9.1, JMP screen.

R:

```
> df1 <- read.table("H:\\Personal Data\\Equivalence &
Noninferiority\\Programs & Output\\d20121025_test_9_1_two_seq_
var_k.csv",header = TRUE,sep = ",")
> df1
     n   sigma_e  sigma_r  beta   del0   dela
1    30   1.665    1.665   0.05   0.33   0.33
2    30   1.665    1.665   0.05   0.33   0.40
3    30   1.665    1.665   0.05   0.33   0.50
4    30   1.665    1.665   0.05   0.33   0.60
5    30   1.665    1.665   0.05   0.33   0.70
6    30   1.665    1.665   0.05   0.33   0.80
7    30   1.665    1.665   0.05   0.33   0.90
8    30   1.665    1.665   0.05   0.33   1.00
9    30   1.665    1.665   0.05   0.33   1.50
10   30   1.665    1.665   0.05   0.33   2.00
11   30   1.665    1.665   0.05   0.33   2.50
12   30   1.665    1.665   0.05   0.33   2.60
13   30   1.665    1.665   0.05   0.33   2.70
```

```
14    30    1.665    1.665    0.05    0.33    2.80
15    30    1.665    1.665    0.05    0.33    2.90
16    30    1.665    1.665    0.05    0.33    3.00
> attach(df1)
> delsig0 <- del0/sqrt(sigma_e**2 ± sigma_r**2)
> delsiga <- dela/sqrt(sigma_e**2 ± sigma_r**2)
> nc0 <- n*delsig0**2
> nca <- n*delsiga**2
> chi_crit <- qchisq(1-beta,n,nc0)
> power <- pchisq(chi_crit,n,nca)
> power
[1]  0.950000000 0.945821364 0.937959697 0.927472866
0.913827934 0.896408739
[7]  0.874544399 0.847556554 0.621395383 0.297415686
0.073417235 0.050355572
[13]  0.033333111 0.021271126 0.013071830 0.007728469
> plot(dela,power)
```

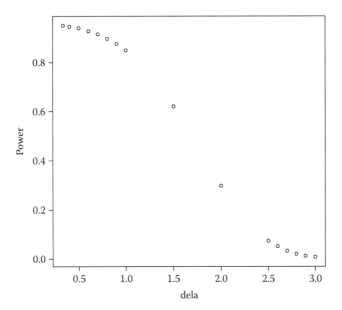

Test 9.1, R: power curve by Δ.

```
> plot(delsiga,power)
```

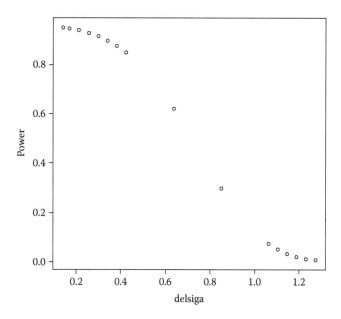

Test 9.1, R: power curve by Δ/σ.

Test 9.2 Two Sequences, Variability Unknown

Input parameters:
  n = sample size (number of points in each sequence)
  g0 = maximum desirable difference, in standard deviation units, between
      the sequences
  ga = alternative differences, in standard deviation units, between the
      sequences
  beta = probability of failure to reject the null hypothesis given g0

SAS code:

```
libname stuff 'H:\Personal Data\Equivalence & Noninferiority\
Programs & Output';

data calc;
  set stuff.d20121025_test_9_2_two_seq_var_u;
  nc0 = n * g0**2;
  nca = n * ga**2;
  f_crit = finv(1-beta,n,n,nc0);
  power = probf(f_crit,n,n,nca);/* Pr{F < = f_crit} */
  run;
```

```
proc print data = calc;/*dataset calc has columns n beta g0 ga
nc0 nca f_crit power */

  run;
```

Obs	n	beta	g0	ga	nc0	nca	f_crit	power
1	30	0.05	0.14015	0.14015	0.58924	0.5892	1.87695	0.95000
2	30	0.05	0.14015	0.15000	0.58924	0.6750	1.87695	0.94923
3	30	0.05	0.14015	0.20000	0.58924	1.2000	1.87695	0.94438
4	30	0.05	0.14015	0.30000	0.58924	2.7000	1.87695	0.92906
5	30	0.05	0.14015	0.40000	0.58924	4.8000	1.87695	0.90404
6	30	0.05	0.14015	0.50000	0.58924	7.5000	1.87695	0.86593
7	30	0.05	0.14015	0.60000	0.58924	10.8000	1.87695	0.81119
8	30	0.05	0.14015	0.70000	0.58924	14.7000	1.87695	0.73723
9	30	0.05	0.14015	0.80000	0.58924	19.2000	1.87695	0.64404
10	30	0.05	0.14015	0.90000	0.58924	24.3000	1.87695	0.53550
11	30	0.05	0.14015	1.00000	0.58924	30.0000	1.87695	0.41959
12	30	0.05	0.14015	1.10000	0.58924	36.3000	1.87695	0.30689
13	30	0.05	0.14015	1.20000	0.58924	43.2000	1.87695	0.20773
14	30	0.05	0.14015	1.30000	0.58924	50.7000	1.87695	0.12914
15	30	0.05	0.14015	1.46000	0.58924	63.9480	1.87695	0.04977
16	30	0.05	0.14015	1.50000	0.58924	67.5000	1.87695	0.03771
17	30	0.05	0.14015	1.60000	0.58924	76.8000	1.87695	0.01754

JMP Data Table and formulas:

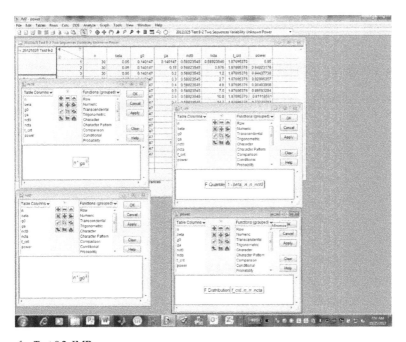

Power for Test 9.2, JMP screen.

R:

```
> df1 <- read.table("H:\\Personal Data\\Equivalence &
Noninferiority\\Programs & Output\\d20121025_test_9_2_two_seq_
var_u.csv",header = TRUE,sep = ",")
> df1
    n beta      g0        ga
1  30 0.05 0.140147 0.140147
2  30 0.05 0.140147 0.150000
3  30 0.05 0.140147 0.200000
4  30 0.05 0.140147 0.300000
5  30 0.05 0.140147 0.400000
6  30 0.05 0.140147 0.500000
7  30 0.05 0.140147 0.600000
8  30 0.05 0.140147 0.700000
9  30 0.05 0.140147 0.800000
10 30 0.05 0.140147 0.900000
11 30 0.05 0.140147 1.000000
12 30 0.05 0.140147 1.100000
13 30 0.05 0.140147 1.200000
14 30 0.05 0.140147 1.300000
15 30 0.05 0.140147 1.460000
16 30 0.05 0.140147 1.500000
17 30 0.05 0.140147 1.600000
> attach(df1)
> nc0 <- n*g0**2
> nca <- n*ga**2
> f_crit <- qf(1-beta,n,n,nc0)
> power <- pf(f_crit,n,n,nca)
> power
 [1] 0.95000000 0.94923176 0.94437738 0.92906357 0.90403908
0.86593284
 [7] 0.81118561 0.73722763 0.64404296 0.53550462 0.41959153
0.30689462
[13] 0.20773002 0.12913863 0.04977451 0.03771117 0.01753920
> plot(ga,power)
```

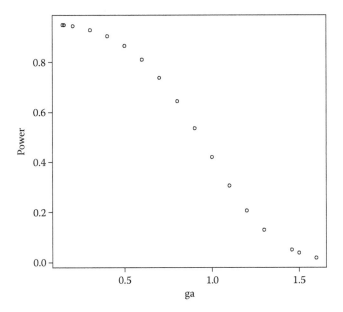

Test 9.2, R: power curve.

References

Anderson, T.W (1958). An Introduction to multivariate statistical Analysis, New York, John Wiley & Sons.

Anderson, S., and W.W. Hauck (1983). A New Procedure for Testing Bioequivalence in Comparative Bioavailability and Other Clinical Trials, *Commun. Statistics Theor. Meth.*, 12(23), 2663–2692.

ANSI/ASQ Z1.9 (1993). Sampling Procedures and Tables for Inspection by Variables for Percent Noncomforming (Milwaukee, WI: American Society for Quality).

Armitage, P. (1971). *Statistical Methods in Medical Research* (Oxford, UK: Blackwell Scientific).

Berger, R.L., and J.C. Hsu (1996). Bioequivalence Trials, Intersection-Union Tests, and Equivalence Confidence Sets, *Statistical Science*, 11(4). 283–319.

Bickel, P.K., and K.J. Doksum (2009). *Mathematical Statistics: Basic Ideas and Selected Topics*, Vol. 1, 2nd ed. (Prentice-Hall).

Bissell, A.F. (1990). How Reliable Is Your Capability Index? *Journal of the Royal Statistical Society (Applied Statistics)*, 39(3): 331–340.

Cowden, D.J. (1957). *Statistical Methods in Quality Control* (Upper Saddle River, NJ: Prentice-Hall).

Desu, M.M., and D. Raghavarao (1990). *Sample Size Methodology* (New York: Academic Press).

Duncan, A.J. (1965). *Quality Control and Industrial Statistics*, 3rd ed. (Richard Irwin Homewood, IL).

Efron, B. (1990). *The Jackknife, the Bootstrap, and Other Resampling Plans* (Philadelphia, PA: Society for Industrial and Applied Mathematics).

Freund, R.A. (1957). Acceptance Control Charts, *Industrial Quality Control*, 14(4): 13–23.

Grant, E.L., and R.S. Leavenworth (1980). *Statistical Quality Control*, 5th ed. (New York: McGraw-Hill).

Greenwood, M. (1926). *The Natural Duration of Cancer*, Reports on Public Health and Medical Subjects, Vol. 33 (London: Her Majesty's Stationary Office), pp. 1–26.

ISO 3951 (1989). *Sampling Procedures and Charts for Inspection by Variables for Percent Noncomforming* (Geneva: International Organization for Standardization).

Johnson, N.L., S. Kotz, and N. Balakrishnan (1995). *Continuous Univariate Distributions*, Vol. 2, 2nd ed. (New York: John Wiley & Sons).

Juran, J.M., and A.B. Godfrey, Eds. (1999). *Juran's Quality Handbook*, 5th ed. (New York: McGraw-Hill).

Kang, C.W., M.S. Lee, J.S. Young, and D.M. Hawkins (2007). A Control Chart for the Coefficient of Variation, *Journal of Quality Technology*, 39(2): 151–158.

Kushler, R.H., and P. Hurley (1992). Confidence Bounds for Capability Indices, *Journal of Quality Technology*, 24(4): 188–195.

Lee, E.T. (1992). *Statistical Methods for Survival Data Analysis*, 2nd ed. (New York: John Wiley & Sons).

Mann, N.R., R.E. Schafer, and N.D. Singpurwalla (1974). (New York: John Wiley & Sons).

Meyer, P.L. (1970). *Introduction to Probability and Statistical Applications*, 2nd ed. (Boston, MA: Addison-Wesley).

Nadarajah, S., and S. Kotz (2007). Statistical Distribution of the Difference of Two Proportions, Letter to Editor, *Statistics in Medicine*, 26(18): 3518–3523.

Schilling, E.G., and D.V. Neubauer (2009). *Acceptance Sampling in Quality Control*, 2nd ed. (Boca Raton, FL: Chapman and Hall/CRC Press).

Schuirmann, D. (1987). A Comparison of the Two One-Sided Tests Procedure and the Power Approach to Assessing the Equivalence of Average Bioavailability, *Pharmacometrics*, Vol.15, No.6, pp. 657–680.

Springer, M.D. (1979). *The Algebra of Random Variables* (New York: John Wiley & Sons).

Welch, B.L. (1947). The Generalization of 'Student's' Problem When Several Different Population Variances Are Involved, *Biometrika*, 34: 28–35.

Wellek, S. (2003). *Testing Statistical Hypotheses of Equivalence* (Boca Raton, FL: Chapman & Hall/CRC Press).

Appendix I: Review of Some Probability and Statistics Concepts

Probability Concepts

Probability begins with the ideas of "sample space" and "experiment." An experiment is the observation of some phenomenon whose result cannot be perfectly predicted a priori. A sample space is the collection of all possible results (called outcomes) from an experiment. Thus, an experiment can be thought of as the observation of a result taken from a sample space. These circular definitions may be a little annoying and somewhat baffling, but they are easily illustrated. If the experiment is to observe which face of a six-sided die lands up after throwing it across a gaming table, then the sample space consists of six elements, namely, the array of one, two, three, four, five, or six dots, as they are typically arrayed on the faces of a six-sided die. Sample spaces need not be so discrete or finite; they can be continuous and infinite, in that they can have an infinite number of outcomes. For example, if a sample space consists of all possible initial voltages generated by LiI batteries made in a battery manufacturing plant, then it would have an infinite (albeit bounded) number of possible outcomes.

A random variable is a mapping from a sample space into (usually) some subset of the real numbers (possible over the entire real line). Think of the random variable as a "measurement" taken after the experiment is performed. Thus, the number of dots in the array showing after the die is cast, or the voltage as measured by a volt meter, would be random variables. There are two basic classes of random variables, discrete and continuous. Discrete random variables are mapped from the sample space to a subset of integers, and continuous random variables are mapped to subsets of real numbers. The die example is discrete, and the voltage example is continuous.

Every random variable has a probability distribution function that describes the chances of observing particular ranges of values for the random variable. In the case of discrete random variables, it also makes sense to talk about the probability of an experiment resulting in a particular value, for example, the probability that the number of dots in the die array showing is four. For continuous variables, it makes sense to talk about the probability of obtaining a value in a "small" range, but the probability of obtaining a particular value is zero. This is not to say that particular values of continuous random variables are never observed or measured; it just means that we do not have the ability to predict a particular value with any non-zero measure of uncertainty.

A probability distribution function describes the probability that a random variable is less than or equal to a particular value. We will use capital letters to represent the random variable, and lowercase letters to represent particular values. If X is a random variable, then the probability function for X is symbolized as

$$F_X(x) = Pr\{X \le x\}.$$

In the case of discrete random variables, this function is a sum of probabilities for particular values, $p(k)$, up to and including the value x:

$$F_X(x) = \sum_{k \le x} p(k).$$

The function $p(k)$ is referred to as the probability mass function. In the case of continuous random variables, the summation is replaced with an integral, and the discrete probability mass function is replaced with something called a probability density function (usually; there are some more or less degenerate cases where a density function does not exist), $f(x)$, which defines the probability that the random variable would have values observed in a small interval, dx:

$$f_X(x)\,dx = Pr\{x - dx \le X \le x + dx\}.$$

So the probability distribution function is:

$$F(x) = \int_{-\infty}^{x} f_X(\xi)\,d\xi.$$

In general, the probability mass functions and density functions are defined in terms of parameters that give these functions their particular characteristics. This book involves several special classes of density functions and their associated parameters.

There are some special characteristics of random variables called moments. We will only be concerned with two such characteristics, called expectation (or mean) and variance (and its square root, called standard deviation). The expectation of a random variable is given by:

$$\mu = \begin{cases} \displaystyle\sum_{k} x_k p(x_k) \\[2em] \displaystyle\int_{-\infty}^{+\infty} \xi f(\xi)\,d\xi \end{cases}.$$

The sum is for discrete random variables, and the integral for continuous random variables. The expectation is like the center of gravity for the random variable, if one thinks about the density function describing the distribution of mass over a beam. The variance is

$$\sigma^2 = \begin{cases} \displaystyle\sum_k (x_k - \mu)^2 \, p(x_k) \\[2em] \displaystyle\int_{-\infty}^{+\infty} (\xi - \mu)^2 \, f(\xi) \, d\xi \end{cases}.$$

Again, the summation and integral are for discrete and continuous random variables, respectively.

Table AI.1 shows the parameteric forms of density and mass functions for several special random variables referred to in the text.

TABLE AI.1

Some Probability Density and Mass Functions

Name	Parameters	Density or Mass Function	Range of Values
Normal	μ, σ	$\dfrac{1}{\sqrt{2\pi}\sigma} \exp\left(-\dfrac{1}{2}\left(\dfrac{x-\mu}{\sigma}\right)^2\right)$	$-\infty < x < +\infty$
Gamma	n, λ	$\dfrac{\lambda^n}{\Gamma(n)} x^{n-1} \exp(-\lambda x)$	$x > 0$
Chi-squared	ν	$\dfrac{(1/2)^{\frac{\nu}{2}-1}}{\Gamma\left(\frac{\nu}{2}\right)} x^{\frac{\nu}{2}-1} \exp\left(-\dfrac{1}{2}x\right)$	$x > 0$
T	ν	$\dfrac{\Gamma\left(\frac{1}{2}(\nu+1)\right)}{\sqrt{\pi\nu}\,\Gamma\left(\frac{1}{2}\nu\right)}\left[1+\dfrac{x^2}{\nu}\right]^{-\frac{(\nu+1)}{2}}$	$-\infty < x < +\infty$
F	ν_1, ν_2	$\dfrac{\Gamma\left(\dfrac{\nu_1+\nu_2}{2}\right)}{\Gamma\left(\frac{1}{2}\nu_1\right)\Gamma\left(\frac{1}{2}\nu_2\right)} \dfrac{x^{\left(\frac{\nu}{2}\right)-1}}{(1+x)^{(\nu_1+\nu_2)/2}}$	$x > 0$
Binomial	n, p	$\dbinom{n}{k} p^k (1-p)^{n-k}$	$k = 0, 1, 2, 3, \ldots, n$

Statistics Concepts

This book is concerned with making inferences about parameters of probability distribution functions. An inference is a generalization made from some specific observations. The specific observations are the data; the generalization is about the values of the parameters. The data are presumed to be a (relatively) small subset of values obtained, measured, or observed in some way from a larger population (sample space). Generally, the parameters are unknown. What we have instead are sample statistics, which are functions of the data. These statistics are themselves random variables, in that every new subset of values from the population yields potentially at least a new value for the statistic. As a result, the sample statistic also has a sample space associated with it, and a probability distribution function as well. The probability distribution function for a sample statistic is often referred to as a sampling distribution function (Meyer, 1970). The form of the sampling distribution usually depends on the formula for the statistic, and the distribution function of the random variable for which the data constitute a subset of values or observations.

One common situation is to make inferences about the expected value of a random variable having a normal probability density, that is,

$$E(X) = \frac{1}{\sqrt{2\pi}\sigma} \int_{-\infty}^{+\infty} x \exp\left(-\frac{1}{2}\left(\frac{x-\mu}{\sigma}\right)^2\right) dx = \mu.$$

It is a convenient coincidence that the expected value of a normally distributed random variable happens to be one of its parameters.

The problem is that both μ and σ, the two parameters for the normal distribution, may not have known values. We can only infer something about this expected value based on a finite subset of values from this normally distributed population. Let $x_1, x_2, x_3, \ldots, x_n$ represent the values of this finite subset, called a sample. We can infer two sample statistics:

$$\bar{x} = \frac{1}{n}\sum_{i=1}^{n} x_i$$

and

$$s = \sqrt{\frac{1}{n-1}\sum_{i=1}^{n}(x_i - \bar{x})^2}.$$

These represent sample estimates for the population parameters μ and σ. An inference would be made if we wanted to know (infer) whether the expected value of this particular normal distribution was equal to a particular value.

If we "hypothesize" about whether the expected value is equal to μ_0, that is, some specific value, we form yet another sample statistic:

$$t = \frac{\sqrt{n}\,(\bar{x} - \mu_0)}{s}.$$

This statistic, if the expected value of the random variable X actually is equal to μ_0, has a sampling distribution called Student's t with a parameter called degrees of freedom (df) equaling the convenient (and known) value $n - 1$. The inference to be made is whether it is believable that the expected value of X is equal to μ_0. If the value of the statistic t falls within a "reasonable" range, we would expect (say, a range that covers 95 percent of values for a random variable having a Student's t-distribution with $df = n - 1$. In other words, if t_p represents the $100p$ percentile of this Student's t-distribution, we would expect the sample statistic to fall somewhere between $t_{0.025}$ and $t_{0.975}$ with probability 0.95 (95 percent). So the inferential rule for this statistic could be the following: If the sample statistic, t, falls in the interval ($t_{0.025}$, $t_{0.975}$), then we are willing to believe that the expected value of the random variable we were sampling is equal to μ_0. Otherwise, we will not believe it.

For an excellent coverage of probability and statistical topics, see Meyer (1970).

Appendix II

Theorem

For any interval on the real line, say, A, and a critical region χ, such that:

$$\sup \Pr \{x \in \chi | \gamma = \inf A\} = 1 - \beta$$

there is always an interval B, such that

$$\inf A > \sup B$$

and

$$\sup \Pr \{x \in \chi | \delta = \sup B\} = \alpha < 1 - \beta.$$

Proof: Let

$$f(\theta) = \Pr \{x \in \chi | \theta\}$$

be a continuous, monotonically decreasing function of θ. Then

$$f(\gamma) = \Pr \{x \in \chi | \gamma\} = 1 - \beta$$

and there is a value $\delta < \gamma$ such that

$$f(\delta) = \Pr \{x \in \chi | \delta\} = \alpha < 1 - \beta$$

by continuity and monotonicity. Let A be any interval such that $\gamma = \inf A$. Since $\delta < \gamma$, define the interval B such that $\delta = \sup B$.

Index

Milton Keynes UK
Ingram Content Group UK Ltd.
UKHW040101071024
449327UK00019B/726